Arm **nce**
Learning

Potential for Reducing Shortages in Army Enlisted Occupations

Michael G. Shanley

Henry A. Leonard

John D. Winkler

Prepared for the United States Army

RAND

Arroyo Center

Approved for public release; distribution unlimited

The research described in this report was sponsored by the United States Army under Contract No. DASW01-96-C-0004.

ISBN: 0-8330-2990-8

RAND is a nonprofit institution that helps improve policy and decisionmaking through research and analysis. RAND® is a registered trademark. RAND's publications do not necessarily reflect the opinions or policies of its research sponsors.

Published 2001 by RAND
1700 Main Street, P.O. Box 2138, Santa Monica, CA 90407-2138
1200 South Hayes Street, Arlington, VA 22202-5050
201 North Craig Street, Suite 102, Pittsburgh, PA 15213
RAND URL: http://www.rand.org/
To order RAND documents or to obtain additional information, contact Distribution Services: Telephone: (310) 451-7002; Fax: (310) 451-6915; Internet: order@rand.org

The Army has established a program to implement distance learning (DL) throughout both its Active and Reserve Component training systems and institutions. At the request of the Army's Deputy Chief of Staff for Personnel (DCSPER), RAND Arroyo Center undertook a research project entitled "Personnel Policy Implications of Army Distance Learning." The goals of this effort were to help the Army maintain readiness and manage personnel efficiently as it implements the various features of The Army Distance Learning Program (TADLP). In the first year of our effort, we presented findings analyzing the implications of TADLP for personnel policy; that analysis documented the need to examine further some of the ways to capitalize on DL to enhance various aspects of readiness.

This report, one of two documenting the results of the research, examines the potential of DL to expedite the Army's efforts to redress personnel shortages in Army enlisted occupations. A companion report (*Enhancing Stability and Professional Development Using Distance Learning*, Henry A. Leonard et al., MR-1317-A, 2001) examines the effect of distance learning on soldier stability and professional development. This research should interest Army and defense policymakers and others responsible for training and human resources development in large, geographically dispersed organizations.

The research was carried out in the Arroyo Center's Manpower and Training Program. The Arroyo Center is a federally funded research and development center sponsored by the United States Army.

For more information on RAND Arroyo Center, contact the Director of Operations (tel 310-393-0411, extension 6500; FAX 310-451-6952; e-mail donnab@rand.org), or visit the Arroyo Center's Web site at http://www.rand.org/organization/ard/.

CONTENTS

Appendix

FIGURES

TABLES

INTRODUCTION

The Army is in the process of implementing The Army Distance Learning Program (TADLP). Its intent is to substitute distance learning (DL) for portions of current resident instruction, with an eye to improving the effectiveness and efficiency of training. Implementation of this program will have a wide-ranging effect on how the Army trains and develops its leaders. It will directly affect how the Army goes about achieving three of its major goals: Manning the Force and Investing in Quality People, Maintaining Unit Readiness and Training, and Training and Leader Development. These are Lines of Operation 3, 4, and 5 specified in the Army's Transformation Campaign Plan.

Since the personnel community plays a key role in achieving these goals, the Deputy Chief of Staff for Personnel (DCSPER) asked RAND Arroyo Center to examine some of the potential implications of DL for personnel readiness. This document provides the results of one portion of that analysis: how DL can help the Army alleviate personnel shortages in the active component among enlisted personnel. Although this research was done for the Army's personnel community, the distance learning program and its implementation are of interest to the Army and the national defense community at large, and not just to those directly concerned with training or personnel management.

Enlisted personnel shortages in the active component are a significant readiness issue documented in a number of reports and Army personnel files, including the Chief of Staff of the Army's (CSA's) monthly readiness reviews. The overall number of shortages in the Army is sizable. First, at the military occupational specialty (MOS) level of detail and distinguishing junior personnel from noncommissioned officers (NCOs), we estimated that the Army was short about 19,300 soldiers in fiscal year 1999 (FY99), representing 5.4 percent of all authorized positions in the enlisted force. About half of these shortages can be traced to a shortfall in total Army personnel relative to authorizations; the other half is created by an imbalance among occupations: overassignment in some occupations and underassignment in others. A second kind of shortage is created when soldiers fill positions for which they lack the proper training. We estimated that about 8,500 E6 and E7 positions, representing about 2.5 percent of all authorizations, were occupied by NCOs not yet formally trained for those jobs or not trained for their grade.

APPROACH

Our analytical approach started by identifying three strategies the Army currently uses to address personnel shortages and that DL could potentially improve: (1) reclassification, (2) cross-training/ MOS consolidation, and (3) acceleration of training. We then chose several cases for more detailed examination in relation to the strategies: MOS 67T (UH-60 Helicopter Repairer), and three MOSs in the Signal area, 31F (Electronic Switching System Operator), 31P (Microwave Systems Operator/Maintainer), and 31U (Signal Support Systems Specialist). Once we completed the analysis of the sample MOSs, we turned our attention to estimating the potential forcewide effects of using DL to address personnel shortage issues.

In assessing how DL-based training strategies might affect shortages, we chose two primary measures of effectiveness (MOEs): changes in shortages (or fill rates), and changes in costs per shortage filled. To support our estimate of the change in shortages, we developed individual inventory projection models (IPMs) for the occupations under study and for the force as a whole. To support our estimate of the change in costs, we developed a suitable methodology based on early military experience with DL courses.

Finally, we note that analyses in this report carry with them the assumption that DL's potential can be fully realized without causing any reduction in the quality of training. Making this assumption allows us to fully explore how much DL might reduce shortages in Army occupations. Moreover, we note that there is a considerable volume of past research supporting the contention that DL, if properly implemented, can provide training as effectively as the classroom training it replaces. However, there can be no doubt that DL's introduction will bring about large and fundamental changes, technically, organizationally, and culturally, in how training (particularly, but not only, institutional training) is conducted in the Army. Viewed in this way, DL clearly poses some risks to the quality of training, especially during the transition period. The keys to maintaining training quality and achieving desired learning outcomes under DL will continue to be careful implementation and monitoring, as well as sustained emphasis and support throughout the Army.

RESULTS

How DL Can Improve the Effectiveness of Reclassification Training

Reclassification of serving soldiers (i.e., the transfer of soldiers from one MOS to another, with necessary retraining provided) is an important means of addressing shortages in certain skills. Historically, the Army reclassifies a sizable number of soldiers each year. In the most recent fiscal year, FY99, the total number of reclassifications was 5,220 soldiers (about 7 percent of the size of total accessions). In addition, an estimated 2,910 prior-service accessions were trained to fill MOS shortages, which is, in essence, another form of reclassification.

One of the reasons the Army uses the reclassification strategy is that it has a number of perceived benefits. First, in reducing shortages, it is more productive for the Army to reclassify a soldier than bring in a new soldier through the accession process. Second, reclassification can be targeted to reduce shortages for Skill Level 1 (SL1) and non-commissioned officers (NCOs) alike. Third, reclassification is more efficient than accession per SL1 shortage filled.

What is the potential value of DL reclassification? First, our analysis suggests that DL could stimulate an expansion of the Army's reclassification program, with a corresponding reduction in the number of MOS shortages. The basis for this expectation boils down to a transaction cost argument. For potential trainees, the transaction costs for entering a new occupation are reduced because DL courses take less time to complete and involve less time away from home and family than the traditional advanced institutional training (AIT) course. For commanders, the transaction costs to retrain surplus unit personnel (to increase unit personnel readiness) are similarly decreased because the training costs less, takes less time to complete, and allows some access to the soldier during the training period. Basic economics argues that lowering the transaction costs of employing the strategy will increase the amount demanded.

While the forcewide benefits of expanding reclassification depend on how large the program can ultimately become, our analysis suggests that gains would be substantial. Enhancing the Army's ability to reutilize existing manpower to meet force structure requirements increases the efficiency of the existing endstrength. One way to value that gain is by the pay and allowances of soldiers moved from surplus to shortage positions: $32,000 per soldier per year for an E4 with 3–4 years of experience. That amounts to a million dollars worth of increased readiness for every 31 additional soldiers voluntarily reclassified.

Second, our analysis suggests that DL could reduce the training cost of reclassification by 30 percent. Assuming an average 10-week AIT course reduced to a 7-week DL reclassification course, we estimate that the training costs (in terms of military personnel and expenditures for operations and maintenance) avoided in filling shortages would amount to $4,500 per additional soldier reclassified, or a million-dollar savings for each additional 222 reclassifications. Moreover, if current reclassifications (and prior-service accessions) could eventually use a DL training mode, we estimate (using the same $4,500 per soldier figure) that savings compared to the current AIT alternative would amount to $29 million per year.

A third benefit of expanding reclassification through DL is the potential opportunity to reallocate selected reenlistment bonuses (SRBs) to increase their effectiveness. We estimate that 5 percent of the 13,500 soldiers receiving bonuses in FY99 were NCOs in occupations with shortages at the SL1 level, but not the NCO level. In such cases, the eventual movement of DL-stimulated SL1 reclassifications to NCO positions could lead to a surplus of NCOs in that occupation.

The surplus, in turn, would allow reallocation of SRBs. Assuming the affected NCOs received the average bonus amount, $6,700, the potential for SRB reallocation due to DL could amount to as much as $4.5 million per year. These funds could then presumably be used to further decrease MOS shortages in the remaining occupations.

How DL Can Improve the Effectiveness of Cross-Training/MOS Consolidation

Cross-training and MOS consolidation are both attempts to produce more effective soldiers, capable of performing a broader range of activities. With cross-training, soldiers already proficient in one MOS are trained to perform related activities in another MOS so that they can informally fill in for that other occupation when necessary. With MOS consolidation, MOSs that perform similar activities are formally combined into one occupation, and soldiers in each of the old MOSs are given additional training to become proficient in all aspects of the new MOS.

Although not employed extensively in the recent past, there is evidence—both from a previous RAND study and from the civilian sector—that cross-training and MOS consolidation represent a feasible strategy when properly implemented. Moreover, it can also be a cost-effective strategy for alleviating the effects of personnel shortages. It accomplishes the task by increasing the skill base of soldiers to make soldiers and the assignment process more effective and efficient. In the case of consolidation, reducing the number of MOSs simplifies the assignment process, reducing force structure imbalances and allowing a reduction in shortages. In the case of cross-training, an increased amount minimizes the impact of personnel shortages by helping redress imbalances in workload. Thus, while the actual number of MOS shortages does not decrease, they are rendered less damaging to readiness.

What is the potential value of using DL to deliver the additional training required for cross-training and MOS consolidation? The answer parallels the effect of DL on the reclassification strategy. First, DL allows for an expansion of the strategy, accompanied by a corresponding reduction of the impact of shortages. The reason is that DL can reduce the transaction costs of training soldiers compared to residential learning (RL) by offering a shorter course closer to home; and lower transaction costs will, in turn, increase the feasibility of using the strategy.

Second, DL could decrease the training cost of cross-training and consolidation. In the case of cross-training, reductions in cost could be even greater than those from DL reclassification. Cross-training, by definition, is dealing with functionally similar MOSs, whereas reclassification might involve two entirely different MOSs. With more overlap between the new and old skills, cross-training could make efficient use of the modular aspect of DL, allowing soldiers to avoid the parts of the reclassification course that cover tasks they have already learned, reducing even further course length, training repetition, and temporary duty (TDY) time.

For MOS consolidation, the way DL would help with training costs depends on how the consolidation is accomplished. If two functionally similar MOSs are simply combined into one (perhaps because of technological change), the use of advanced learning technology might contribute to the development of a feasible training strategy by reducing the time required to complete it. But if the concept is to produce a generic specialist across two or more specialties (as is true for helicopter repairers in the civilian world), DL could provide much or all of the equipment-specific training soldiers need for a specific assignment without leaving their home station.

From a forcewide perspective, the potentially large increase in MOS consolidation in the near future suggests that this strategy can have a large impact in avoiding future training costs. For example, the organization of future aviation brigades suggests compatibility with MOS cross-training and MOS consolidation. More broadly, proponents working with the Army Development System (ADS) XXI Task Force have submitted a list of 44 MOSs for consolidation, involving 17 percent of authorizations. In addition, this task force will recommend future consolidation, in as many as 88 more MOSs beyond the 44 already submitted.

How DL Can Improve the Effectiveness of Accelerating Training

When NCOs cannot be trained in a timely manner, the result is an increase in the shortage of trained personnel. For FY99, we estimate that 8,500 E6 and E7 positions were occupied by soldiers not formally trained (with the appropriate Basic NCO Course (BNCOC) or Advanced NCO Course (ANCOC)) for those jobs or not trained for their grade. That number represents 2.4 percent of all authorizations, and 8.9 percent of E6 and E7 authorizations. Some of the untrained are NCOs not yet formally trained for their grade, but most are "fast-trackers" serving above their grade in positions for which there would otherwise be a shortage.

With proper support and monitoring, DL could make BNCOC and ANCOC training possible earlier in the select-train-promote sequence. First, DL training can begin before scheduled residence training courses are available. Second, DL training can be taken in small pieces, on a "continuous" basis. Third, DL training can occur at home station. Fourth, modularized DL courses allow "testing out" of already mastered material, which means fast-trackers who get much of their experience through on-the-job training (OJT) would not have to sit through the parts of course material they have already mastered. Finally, DL can enhance the type of self-development training that can accelerate the institutional training process.

Given the extent of training shortages among NCOs, we find the forcewide potential for DL to reduce shortages by accelerating training relatively high. Moreover, with the new NCO Educational System (NCOES) model projecting more individual training for NCOs, we think the level of potential application of DL for BNCOC and ANCOC will increase in the future, working to avoid an increase in future training shortages.

CONCLUSIONS

Using DL in the contexts described above offers significant potential for further reducing shortages and for reducing the marginal cost of achieving those reductions. Realizing that potential, however, requires implementing the DL program in ways that will produce the available benefits. This means early selection of courses for conversion that will do the most to reduce the shortage problem (i.e., courses, especially longer courses, focused on shortage MOSs, consolidating MOSs, and MOSs with ANCOC and BNCOC backlogs). Most important, it means creating DL courses that are attractive to students, commanders, and the Army, courses with sufficient flexibility to be easily integrated into varying soldier career paths. In this regard, the DL program should emphasize maximum use of emerging learning technologies to help reduce learning time (thus shortening course time) and to allow significant portions of the training to be completed at home station. In addition, the DL program should strive to avoid pitfalls found in the past in industry and academia. This means providing sufficient student support to ensure the speedy completion without increased personnel tempo (PERSTEMPO) or course attrition and providing sufficient administrative support for scheduling, monitoring, and recording training results. Finally, DL needs to provide courses as modularized, "just-in-time" training to take full advantage of opportunities to reduce unnecessary training and to allow refresher training on demand.

The above list of specific DL characteristics for achieving the Army's personnel readiness goals underscores the need for DCSPER, as well as the Army as a whole, to work closely with the training community to develop the kind of DL program that can maximize benefits in all parts of the Army.

ACKNOWLEDGMENTS

We are indebted to many officials from Headquarters, Department of the Army, from TRADOC and its installations, and from other supporting agencies. In particular we would like to thank Maria Winston, Jim Coats, LTC Norvel Dillard, and the rest of the staff of the Training Requirements Division in the DCSPER's Directorate of Military Personnel Management. LTC Eli Alford and MAJ Gene Piskator of the Army's Personnel Command provided a wealth of useful data and helpful advice. We were greatly aided in our understanding of Armor Branch personnel structure and the Armor Center's distance learning initiatives by COL Richard Geier, Connie Wardell, Bob Bauer, Aubrey Henley, Leonard Fizer, CPT Chad Jones, and George Paschetto. Similar valuable assistance came from the Army Signal Center's COL Larry Turgeon, SGM Dale Manion, Rosemary Berlin, Lilla Dancy, Howard Moore, Renee Carmichael, and Kelly Larsen. LTC Joe Charsagua of TRADOC was invaluable in helping us sift through the wealth of data we had been provided by TRADOC and the Army in support of this project. Angela Lewis and John Bryant of the Program Management Office for TADLP provided helpful insights into the complex process of cost and savings estimates. Finally, we are indebted to our colleagues Charles Kaylor and Herb Shukiar for their insights and comments on earlier drafts of this report.

AC	Active Component
ACES	Army Continuing Education System
ADS	Army Development System
AFQT	Armed Forces Qualifying Test
AIT	Advanced Individual Training
ANCOC	Advanced NCO Course
ASI	Additional Skill Identifier
ASMP	Army Strategic Management Plan
AT	Annual Training
ATRRS	Army Training Requirements and Resources System
BNCOC	Basic NCO Course
BOS	Base Operating Support
CMF	Career Management Field
CONUS	Continental United States
CSA	Chief of Staff of the Army
DCSOPS	Deputy Chief of Staff for Operations and Plans
DCSPER	Deputy Chief of Staff for Personnel
DL	Distance Learning

DoD	Department of Defense
EMF	Enlisted Master File
FY	Fiscal Year
IET	Individual Entry Training
IPM	Inventory Projection Model
IPR	In-Process Review
MOE	Measure of Effectiveness
MOS	Military Occupational Specialty
MOSLS	MOS Level System
NCO	Noncommissioned Officer
NCOES	NCO Educational System
NGB	U.S. Army National Guard Bureau
OJT	On-the-Job Training
O&M	Operations and Maintenance
PCS	Permanent Change of Station
PERSCOM	U.S. Army Personnel Command
PERSTEMPO	Personnel Tempo
PMAD	Personnel Management Authorization Document
QM	Quartermaster
R&D	Research and Development
RC	Reserve Components
RL	Residential Learning
RMC	Regular Military Compensation
SKAs	Skills, Knowledges, and Abilities
SL	Skill Level
SRB	Selective Reenlistment Bonus
TADLP	Total Army Distance Learning Program

TASS	Total Army School System
TDA	Table of Distribution and Allowances
TDY	Temporary Duty
TOE	Table of Organization and Equipment
TRADOC	U.S. Army Training and Doctrine Command
TTHS	Trainees, Transients, Holdees, and Students
VTC	Video Telecommunications Conference

Chapter One

INTRODUCTION

BACKGROUND

The Army has established The Army Distance Learning Program (TADLP) under the auspices of the U.S. Army Training and Doctrine Command (TRADOC). The intent of this program is to capitalize on the capabilities of distance learning (DL) technology to replace resident instruction with DL in those cases where the technique is suitable to teach the material. In effect, this means dividing existing courses into resident learning (RL) and DL phases or modules. Thus, TADLP will significantly change how individual training is conducted—how leaders and soldiers are developed—both in institutions and in the field.

The Army's investment in distance learning amounts to about $630 million, covering infrastructure, expenses involved in developing courseware, fielding costs, and the management costs tied to program development and implementation. These costs have been estimated through the year 2015, but most of them ($440 million) are in the early and middle stages of that period.[1] The infrastructure investment will provide networks and hardware (e.g., classrooms,

---

[1]The source for these figures is TADLP's *Economic Analysis,* published by the Program Management Office, TADLP, September 2000. In addition, the Army National Guard's Distributed Training Program had about $220 million in investment and operating costs through FY00 that are not included in these figures. The Army also has other programs, currently outside the purview of TADLP, that are using or will be able to capitalize on DL technologies. These include computer-based training, DL support for the Army's Continuing Education Program, and Army University Access Online, the new initiative to provide greater access to college courses.

computer workstations) for DL sites. As of April 1999, TADLP planned to support 844 DL sites in 454 locations in the continental United States (CONUS) and abroad. The investment in courseware provides for converting portions of 525 courses to DL.

The Army is pursuing these changes because it believes a number of benefits accrue from DL. These benefits amount to direct or indirect enhancements to training and personnel readiness. DL creates a potential for delivering targeted training on short notice, can facilitate access to education, and may provide more timely training than a resident course. And because technology can enhance the speed of learning, course lengths can decrease and soldiers may spend less time away from their units and less time between operational assignments. Finally, some resource savings may also be possible under DL. These could take the form of reduced travel costs and reductions in personnel resources devoted to the delivery of institutional training, allowing endstrength to be shifted from TDA to TOE organizations.

Several key features of the DL program determine how it will affect training, the soldiers and leaders being trained, and their units. First, DL moves a significant portion of training from the traditional schoolhouse into locations near where the soldiers reside, making it easier for them to attend. Second, DL offers, in lieu of traditional schoolhouse resources, emerging educational technology and media to provide increased access to training material and to deliver the training. Third, by not requiring soldiers to leave their units for RL courses elsewhere and by providing significant amounts of training in asynchronous (i.e., self-paced) modes, DL provides the potential for increasing flexibility and continuity in the timing of training. Finally, because it moves some training out of directly supervised classrooms and school environments, DL increases the responsibility of soldiers and their chain of command for ensuring timely completion of training.

While the distance learning program is under the purview of the Army's training community (i.e., primarily TRADOC and the Army's Deputy Chief of Staff for Operations and Plans (DCSOPS)), the program has broader implications for the Army as a whole. TADLP will directly affect the ways the Army will achieve three of its overall strategic goals: training, quality people, and leader development. All three concern the Army at large; two of them—quality people and leader development—are a primary responsibility of the Army's Deputy Chief of Staff for Personnel (DCSPER). Because of these wide-reaching potential effects, it is not just the training community, but also the Army as a whole and DCSPER in particular, that have a large stake in the development of the distance learning program and the direction it takes.

The personnel implications of TADLP (and thus a significant part of the Army's stake in the program) boil down to readiness: can TADLP help to enhance the personnel readiness of the Army? Many of the features of DL—chief among them shorter overall training time, the availability of "on-demand" training packages, and greater flexibility in scheduling—can enhance personnel readiness if judiciously employed.

To examine DL's potential effects on readiness, we look at personnel readiness at three levels: Army-wide, organizational, and individual (see Table 1.1). Army-wide personnel readiness depends on the overall natural abilities, training and education, and morale of the Army's people (these are also components of individual readiness) and on the Army's ability to develop, train, position, and motivate those people to accomplish their assigned missions. Organizational readiness includes the above considerations, and it looks more specifically at the degree to which the skills and qualifications of the soldiers in units and organizations match the skill and qualification requirements specified for those units and organizations. Of the three forms of readiness, this is the easiest to quantify: improving the match between the skills of the soldier inventory and the requirements of the organization improves organizational readiness.[2] Individual readiness—the skills, training level, general aptitude, and motivation/morale of each individual—is the foundation for the two collective forms of readiness.

DL could potentially contribute to readiness at all three levels. The right-hand side of Table 1.1 specifies how DL could help in each area. Perhaps the key words in the table are "could help." Aside from the concern over whether DL as a tool can actually deliver what its pro-

---

[2]The Army uses statistical measures of this match as part of its unit readiness assessments.

ponents promise—a legitimate concern that we address in Chapter Three—there is also the concern about what empirical evidence exists to support the claims that DL can help accomplish the bulleted items in the table. It is this concern that led to our research focus, a subject we turn to next.

PURPOSE OF THIS REPORT

We undertook empirical analyses for the DCSPER to determine how much DL can help improve personnel readiness in the three areas shown above. Our findings should be helpful both to the personnel community and to the Army at large in evaluating DL's potential and

Table 1.1

How DL Could Help Address Three Levels of Readiness

Readiness Level	Definition	How DL Could Help
Army	The degree to which the Army is able to develop, train, position, and motivate its personnel to accomplish their assigned mission.	• Enable increased course enrollment, graduation rates • Enhance ASI, other functional training • Speed promotion qualification • Reclassify, cross-train, consolidate MOSs
Organizational	The degree to which soldiers' skills and qualifications match the requirements of their units and organizations. Measured through the Unit Status Reporting System.	Above plus: • Shorten formal training time • Decrease time away from home • Increase available days to the unit • Enable improved mobilization processes • Provide refresher/new equipment training
Individual	The skills, training level, general aptitude, and motivation and morale of each individual.	Above plus: • Enrich leader development • Expand opportunities for personal and professional development

ways to capitalize on it. This report and a companion report present the results of our research.[3]

In this report we examine how DL can help the Army more quickly address active component manpower shortages in understrength skills. We look at DL's potential to enable faster completion of re-classification training, faster NCO promotion qualification, and more efficient forms of additional skill training. Success in these areas would improve the skill mix component of the Army's overall readi-ness posture[4] and in turn also improve the skill content in units and organizations, enhancing organizational readiness.

Of course, the effect on organizational readiness also depends on the judicious distribution of the additional trained soldiers into units and organizations where there are shortfalls. Thus, we find that DL enables, but does not guarantee, better organizational readiness. A common theme in our research is that DL can serve as an enabler for certain institutional strategies that would be undesirable or infeasi-ble in the absence of DL.

The companion research report (Leonard et al., 2001) takes a closer look at what DL programs might do to reduce the time soldiers spend away from both unit duties and their families, improving organiza-tional readiness by enhancing stability in units and quality of life for soldiers and families. That report also describes some of the ways the DL program could help overall individual readiness: not only by improving skill qualifications and quality of life, but also by enriching leader development and expanding other opportunities for personal and professional development.

ORGANIZATION OF THIS DOCUMENT

The following chapters discuss how DL can help the Army alleviate personnel shortages among its enlisted personnel, starting in Chap-ter Two with a brief discussion of why personnel shortages are a readiness problem. Chapter Three discusses the strategies the Army

---

[3]The companion report is Henry A. Leonard et al., *Enhancing Stability and Professional Development Using Distance Learning*, Santa Monica, CA: RAND, MR-1317-A, 2001.

[4]That is, bring the manpower fill in each skill area closer to requirements.

currently uses for alleviating these shortages and identifies areas where DL can be of help. This chapter also documents the approach we use to determine how useful DL could be. Chapters Four, Five, and Six describe the results of applying the approach to determine the effectiveness of DL-based approaches for three Army strategies to alleviate personnel shortages. Chapter Seven offers some general conclusions and next steps.

Chapter Two

THE PROBLEM OF PERSONNEL SHORTAGES FOR READINESS

Throughout the discussion here and in the remainder of the report, we define a "personnel shortage" as a situation where trained and available assignments (personnel ready to fill a position)[1] fall short of authorizations (the number of positions established by the Army for a particular MOS). Such shortages indicate a readiness problem for the Army in that authorizations are the best direct statement of personnel and skills required for readiness. In other words, shortages—the gaps between authorizations and assignments—measure a direct detriment to personnel readiness.

Although in subsequent chapters we discuss the Army's strategies to address personnel shortages and, more important for this research, DL's potential to enhance those strategies, here we take a step back and briefly discuss the scope and sources of personnel shortages.

WHAT IS THE SCOPE OF THE PERSONNEL SHORTAGE PROBLEM IN THE ARMY?

Shortages in personnel can be divided into four categories useful for the analysis described in this report. First, shortages occur when the Army does not attract enough personnel, indicating a recruiting shortfall or lower-than-needed retention rates for those already recruited. Second, the distribution process produces shortages when some occupations are assigned more personnel than authorized,

---

[1]Assignments are endstrength less soldiers in the Trainees, Transients, Holdees, and Students (TTHS) accounts.

creating an offsetting shortage in other occupations. This type of shortage can be the result of Army distribution decisions, of force structure changes,[2] or of intentional policy.[3] Third, the distribution and training processes lead to shortages when soldiers are assigned to positions but are not trained for them. In other words, soldiers are filling some positions but not of the "right type." Fourth, shortages occur when soldiers are both assigned and trained for their positions but are not available for deployment.

The overall number of shortages in the Army is sizable and larger in some areas than in others. We discuss here those categories of highest relevance to our analysis of DL.[4] First, at the MOS level of detail and distinguishing junior personnel from noncommissioned officers (NCOs), we estimated that the Army was short about 19,300 soldiers in fiscal year 1999 (FY99), representing 5.4 percent of all authorizations.[5] About half of these shortages can be traced to a

---

[2]An imbalance in the force can be created not only by a change in the number of available soldiers, but also by a change in the number of authorized positions. Thus, personnel managers who successfully achieve a desired fill rate in an occupation can suddenly have a shortage or surplus created by a short-term change in the number of force structure authorizations.

[3]The Army has, at times, intentionally created shortages and surpluses; for example, combat MOSs might be intentionally overaccessed relative to other MOSs to ensure that personnel fill is achieved in those occupations.

[4]For example, we do not attempt to account for the number of soldiers who cannot deploy in wartime. However, that number has recently been estimated by RAND researchers at 3.5–4 percent of several large Army units and installations. See J. Michael Polich, Bruce R. Orvis, and W. Michael Hix, *Small Deployments, Big Problems,* Santa Monica, CA: RAND, IP-197, 2000.

[5]Personnel shortage estimates were derived for FY99 by taking the difference between the monthly averages of authorizations and operating strengths in FY99. Authorizations came from the Personnel Management Authorization Document (PMAD), while operating strengths were derived from the Enlisted Master File (EMF), calculated by subtracting those in the TTHS accounts from total Army endstrength. Using those two files, shortages could be calculated down to the MOS and grade level of detail.

To validate our findings, we compared the monthly shortages we calculated with those reported by the Active Army Military Manpower Program and given to us by Headquarters, Department of the Army, Office of the DCSPER, Military Strength Analysis and Forecasting Division. We found that our computation of FY99 shortages was about 15 percent higher than theirs, primarily because of the higher number we calculated for TTHS. The Army has been undergoing some change in how it computes TTHS, and it performs some offline adjustments to the numbers that we were able to document only partially.

shortage in total personnel, while the other half are created by surpluses in other Army occupations. The problem was worse for junior personnel, where personnel shortages amounted to 7.5 percent of all authorizations, than for NCOs, where personnel shortages amounted to 3.1 percent of all authorizations.

Second, the number of untrained Army personnel added to the shortage problem. While nearly all AC soldiers occupying junior-level positions were trained for those positions, we estimated that in FY99 about 8,500 E6 and E7 positions, representing about 2.4 percent of all authorizations, were occupied by soldiers not formally trained for those jobs or not formally trained for that grade.[6] In about a third of these cases, soldiers had not yet received the training appropriate for their grade, that is, the Basic NCO Course (BNCOC) or Advanced NCO Course (ANCOC). In the other two-thirds of the cases, soldiers were serving in a position above their grade level without having completed its required formal training. Specifically, this means that either E5s were serving in E6 positions without the appropriate BNCOC course, or E6s were serving in E7 positions without the appropriate ANCOC course.

HOW ARE MOSs WITH A PERSONNEL SHORTAGE PROBLEM IDENTIFIED?

Given that such shortages exist, where do we see them in the Army? They are documented in a number of reports and Army personnel files, including the "Critical MOS List" (a part of the Chief of Staff of the Army's (CSA's) monthly readiness reports), lists of low-retention MOSs generated from the Enlisted Master File (EMF), and the U.S. Army Personnel Command's (PERSCOM's) monthly MOS Data Sheets. Below we discuss these three sources in more detail.

Critical MOS List

The primary way to locate potential personnel shortages is to consult the CSA's monthly readiness reports. The reports not only list the

---

[6]The source of this information is the EMF and ATRRS. We did not try to enumerate training shortages at other NCO grades because, as we shall discuss in future chapters, that training is less applicable to our analysis.

amount of the shortage, but also often identify a cause and the status of ongoing efforts to eliminate the shortage, along with a projected "get well" date. Chronic shortages are the most problematic for the Army, because they indicate that prolonged effort using existing strategies has not been able to solve the problem. Table 2.1 lists the top ten MOSs with the most chronic shortages in the period between the beginning of FY97 and the beginning of FY00. The metric used to rank each MOS is the number of times it has appeared on the list. Note that four MOSs made the list in all 36 months of the period of interest. All the occupations identified in the table appeared on the list at least 21 out of 36 months. An additional five MOSs appeared at least half of the months, and five more at least 12 of 36 months. In total, 20 MOSs appeared on the list a third of the time or more. Those 20 account for 22 percent of total authorizations in the force.

Lists of Low-Retention MOSs

A second and more indirect method for identifying occupations that may have shortages is to focus on those with low retention rates.

Table 2.1

Identifying Personnel Shortages Using the Critical MOS List

MOS	Title	Branch	Months on Critical MOS List FY97–00	FY99 Authorization
67T	UH-60 Helicopter Repairer	Aviation	36	3,367
92Y	Unit Supply Specialist	Quartermaster	36	11,739
96B	Intelligence Analyst	Intel.E5	36	3,009
98G	Voice Interceptor	Signal	36	2,495
54B	Chemical Operations Specialist	Chemical	34	5,933
31F	Electronic Switching System Operator	Signal	33	3,511
31S	Sat Com System Operator/Maintainer	Signal	31	1,593
77F	Petroleum Supply Specialist	Quartermaster	31	7,600
45E	M1 Abrams Tank Turret Mechanic	Armor	23	550
45T	M2 Bradley FV System Turret Mechanic	Armor	21	412

SOURCE: CSA's Critical MOS List, October 1996 through October 1999, and the PMAD.

Occupations with low retention rates are likely to also produce shortages at the affected grade levels. We identified low-retention MOSs from snapshots of the Army's Enlisted Master File (EMF). Using the October 1997 and October 1998 EMF, low-retention-rate MOSs were identified by calculating, for each MOS, the percentage of personnel in the inventory at the beginning date who were still in the force one year later, then ranking the MOS with the lowest percentages first. Table 2.2 shows the low-retention-rate MOS list for the E5 and E6 grades in the AC force. The list only includes MOSs with 200 or more in the appropriate grades.[7]

Table 2.2

Identifying Personnel Shortages Using Low-Retention MOSs

MOS	Title	October 1997 Inventory	1997–1998 Retention
Grade E5			
98C	Signals Intelligence Analyst	245	47%
71L	Administrative Specialist	732	51%
97B	Counterintelligence Agent	257	55%
75B	Personnel Administrative	494	56%
35E	Radio and Communications	363	57%
91C	Practical Nurse	450	60%
Grade E6			
71L	Administrative Specialist	248	45%
31R	Multichannel Transmission Systems Operator	387	54%
97B	Counterintelligence Agent	205	57%
11B	Infantryman	2,009	60%
31U	Signal Support Systems Specialist	477	62%

SOURCE: EMF.

---

[7]We checked the retention statistics in Table 2.2 against those presented in the FY98 PERSCOM Retention Report. Although complicated by the fact that the PERSCOM report categorizes retention by soldier life cycle (i.e., first term, mid-career, career) rather than by grade, we found that the reports generally agreed, with a couple of exceptions. Given that Table 2.2 has illustrative purposes only in our analysis, we did not attempt to reconcile the few existing differences.

Sample MOS Data Sheet Comparisons by Authorizations and Assignments

Finally, the MOS data sheets produced monthly by PERSCOM offer a way to identify shortages one occupation at a time. Examining detailed authorizations and assignments for specific MOSs allows us to see not only the size of the shortage but also where it occurs in the force—that is, at the entry level (skill level (SL) 1, grades E3–E4), or at the NCO level (grades E5, E6, E7, E8, or E9). Figure 2.1 shows two examples of shortages that occur in the junior ranks, one for MOS 67T (which is atop the list of chronic critical MOSs shown in Table 2.1) and one for MOS 31U, which landed on the Critical MOS List five times between October 1996 and October 1999. In both cases, we see that shortages occur at the E3–E4 level. The columns represent the assigned personnel for the grade, while the lines represent the number of personnel authorized at that grade. The shaded cross-hatched portion represents the shortage in the fill rate, which is 86

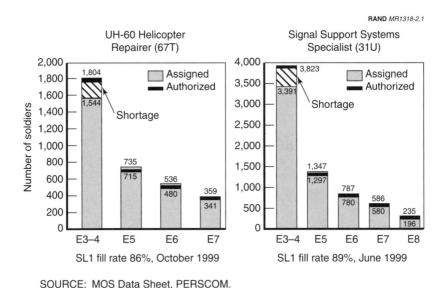

SOURCE: MOS Data Sheet, PERSCOM.

Figure 2.1—Some Personnel Shortages Occur at Entry Level (SL1)

percent for 67T (in October 1999) and 89 percent for 31U (in June 1999). Note that in neither case were the shortages in SL1 matched by shortages at the NCO level. Often, the number of NCOs can be kept to authorized levels by the management of promotion and the use of selective reenlistment bonuses (SRBs).

As noted above, in some cases shortages can also be seen at the NCO level. Figure 2.2 shows two example MOSs at this level, both from the Signal Branch: 31F (which is one of the chronic critical MOSs shown in Table 2.1) and 31P, a relative newcomer to the critical list. Notice that the shortages occur in different parts of the NCO ranks. In particular, 31F shows an 88 percent fill rate at the E5 level (as of June 1999), while the 31P shows a 76 percent fill rate at the E7 level (as of June 1999). Often, the optimum strategy for eliminating a shortage will differ depending on the grade level at which it appears.

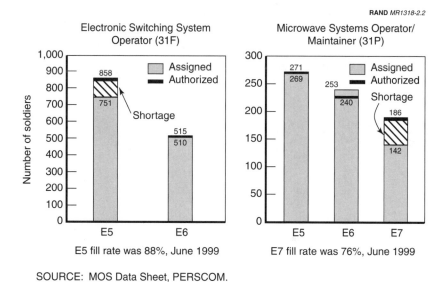

SOURCE: MOS Data Sheet, PERSCOM.

Figure 2.2—Other MOSs Have Shortages at the NCO Level

AN APPROACH TO ANALYZING APPROPRIATE DL-BASED ARMY STRATEGIES FOR ALLEVIATING PERSONNEL SHORTAGES

Given that personnel shortages pose readiness problems for the Army, the next question is whether DL can help in alleviating those shortages and thereby help improve readiness. The idea that DL might have an impact on alleviating shortage MOSs is based on DL's potential for being faster and more efficient than traditional residential training. For example, DL can give students more access to training time and materials. As a result, training might be designed to begin immediately after a requirement is identified and at or near home station, rather than having the soldier wait for an opening in a resident training program at a school or training center—a class that would require temporary duty (TDY) or permanent change of station (PCS) orders. Further, access to DL is improved if training lessons can be broken into smaller parts, more suitable for working into the continuing responsibilities of an existing job. Also, DL can improve training focus. If courses are modularized into distinguishable parts, students can take only that part of the training that is critical to their assignment. Finally, as technology can enhance the speed of learning, DL course lengths can decrease.

Of course, all these benefits speak to the "potential" of DL, because DL's potential really has not been quantified. In this chapter, we outline an approach to help us understand the real benefits behind the potential. The first task in this approach was to summarize strategies that the Army already uses for reducing MOS shortages and look to see how DL could be an enabler and expediter for those strategies. Second, we chose specific MOS examples to measure how DL might help lessen either the shortage itself or the cost of reducing it. In defining how DL would become part of existing strategies, we did not tie ourselves to current Army DL training patterns or course characteristics. Rather, we assumed that a range of DL course characteristics would be available in the future. Third, we built an inventory projection model (IPM) to estimate the long-term effects on inventories of DL-based and non-DL-based strategies. Fourth, we constructed some measures of effectiveness (MOEs), such as higher fill rate (or smaller shortage) and lower cost for the same fill rate, to use to evaluate DL's effects. Finally, we sought to extrapolate our findings—when possible—beyond the MOSs directly examined to gain some insight into forcewide effects.

STRATEGIES TO REDUCE SHORTAGE MOSs AND HOW DL COULD HELP

The fact that there is an ongoing problem with shortage MOSs, which was discussed in Chapter Two, is not news to the Army. In fact, the Army regularly takes active steps to reduce shortages. Four generic strategies for alleviating shortages are summarized in Table 3.1.[1]

The first and last strategies are basically recruitment and retention, respectively, and address the problem by changing the relationship with the outside environment: Get more soldiers in at the beginning of the process and keep more soldiers from dropping out at the end of the process. The other two strategies—expand training (with its four substrategies) and consolidate MOSs—alter internal Army processes to address shortages caused by inventory imbalances or training logjams.

Given these four strategies, how could DL be used to help the Army deal with the shortage MOSs problem? Table 3.2 presents seven hypotheses of how DL might help. DL might increase accessions if the existence of the program, equipment, and facilities offered more access to educational opportunities (civilian and military) through the World Wide Web or other DL media; in that case the Army could look more attractive to potential enlistees as a place to receive useful training and education. For similar reasons, the Army might look

---

[1]Although not discussed here, the Army has a number of force structure options available to reduce shortages; for example, shortages could be reduced by outsourcing an occupation's function or by redesigning unit structures (through TDAs and TOEs).

Table 3.1

Generic Strategies for Helping Reduce Shortage MOSs

Strategy	Description
Increase accessions	• Bring in new recruits to fill shortage MOSs
Expand training	
Increase quotas	• Increase the number of training seats for shortage MOSs
Reclassify	• Reclassify soldiers to shortage MOSs
Cross-train	• Increase soldiers for shortage MOSs by cross-training those in similar MOSs
Accelerate training	• Fill shortage MOSs in higher grades by training surplus soldiers in lower grades faster
Consolidate MOSs	• Increase soldiers available for shortage MOSs by consolidating similar MOSs
Increase retention	• Retain existing soldiers in shortage MOSs

Table 3.2

How DL Might Help Reduce Shortage MOSs

Strategy	How DL Could Help
Increase accessions	• Provide expanded educational and training opportunities
Expand training	
Increase quotas/ enrollments	• Allow for more flexible training at or near home station
Reclassify	• **Use in reclassification course**
Cross-train	• **Use in reclassification module**
Accelerate training	• **Use in BNCOC and ANCOC**
Consolidate MOSs	• **Use in specialized training**
Increase retention	• Provide expanded educational and training opportunities

more attractive to those considering reenlistment. In fact, the Army is currently using DL-based strategies for accession and retention, although it is not clear that DL is being used to deal specifically with the shortage MOSs problem.[2]

Under the "expand training" category of strategies, DL could help reduce shortages by improving the distribution of existing endstrength (moving personnel from surplus to shortage areas) and, indirectly, by helping the Army recruit and retain more soldiers. (This element will be discussed more fully below.) Reclassification courses, or parts thereof, are the vehicle for reclassification and cross-training, and these could be expanded under DL. The BNCOCs and the ANCOCs are the potential vehicles for training acceleration. In the case of consolidation, the DL vehicle is specialization training, which is the training that soldiers in a more generic MOS would take to receive an assignment involving specific equipment.

The four entries shown in bold type in Table 3.2 are the ones we address in this report. We judged that the other three strategies would require a more fully implemented DL program before analysis could effectively test their effects on shortages.

SELECTION CRITERIA FOR CHOOSING MOSs FOR FURTHER STUDY

After defining the DL-based strategies for reducing shortage MOSs and the four strategies we decided to focus on, we chose a number of MOSs for more in-depth analysis. Cases for further study were chosen using five criteria. First, we sought occupations with a history of large shortages or low retention (as illustrated earlier in Chapter Two). Second, we sought occupations that are training-intensive, that is, the MOSs have required training periods that are at or longer than the average. Since in DL we are analyzing the potential impact of a training intervention, the longer the required training period, the greater the potential for a large impact. Third, we sought MOSs with a sizable portion of their population overseas, where the potential impact of DL could be the greatest. Fourth, we wanted to deal with

---

[2]See "Soldier by Day, Online Student by Night," *Los Angeles Times*, Monday, July 10, 2000, p. A1.

U.S. Army Training and Doctrine Command (TRADOC) schools that have some experience with DL. Fifth, we wanted to steer clear of MOSs that might become less important or critical to the Army in the future. This could happen either because the skills themselves are expected to be less critical in the future or because the skills, while still critical, are not expected to require more training (as most MOSs are) as the technological sophistication of the Army increases.

These criteria led us to choose the UH-60 Helicopter Repairer MOS (67T) and several MOSs in the Signal area—Electronic Switching System Operator (31F), Microwave Systems Operator/Maintainer (31P), and Signal Support Systems Specialist (31U)—as our candidates for further study.

MEASURING THE EFFECTS OF DL-BASED AND NON-DL-BASED STRATEGIES ON SHORTAGES

To support the analysis of how DL- and non-DL-based strategies (especially reclassification) in these cases affect the size of the shortage, we developed individual Inventory Projection Models (IPMs) for the occupations under study. These IPMs were designed to measure the long-term impact of the DL-based strategies on the size of the personnel inventory, and thus on the fill rate and the size of the shortage. Required data on promotions and losses and continuation rates were obtained from Army personnel files showing the distribution of existing personnel and the expected distribution of personnel in the future. Appendix A discusses the design of these models in more detail, and it provides a conceptual description of how the models work.

MEASURING THE EFFECTS OF STRATEGIES ON THE COST OF REDUCING THE SHORTAGE

In addition to measuring the effect of DL- and non-DL-based strategies on the size of the shortage, we measured the effect on cost. To make the required comparisons for our selected MOSs, we estimated the cost of accession, reenlistment, reclassification with traditional RL training, reclassification with DL training, cross-training with traditional RL training, and cross-training with DL training. The major cost elements in this effort were pay and allowances of students, the cost of bonuses, and the aggregated cost of required training.[3]

To make the comparisons between DL- and non-DL-based strategies, we had to compare the relative costs of operating DL and RL training courses. The largest component of the recurring cost of DL, and the one of most concern to DCSPER, is personnel costs. While DL will clearly reduce some personnel requirements (e.g., there will be less need for instructor platform time), it will also introduce other requirements (e.g., the need for instructors to assist and monitor DL students). Even in cases where DL is a good substitute for RL, it will still generate requirements for instructors and support personnel. Instructors will be needed to conduct synchronous classes as required, monitor student progress and provide student feedback, keep courseware current, and provide quality assurance. Support personnel will need to provide administrative support, maintain Web sites and software, and help in keeping courseware current. Although these people can be the same ones who perform similar functions for residential training, resource managers must consider that the functions have to be performed for both RL and DL segments and, thus, that the schools and centers have to be staffed with this entire workload in mind.

There is evidence from documented commercial and academic experience that DL can be more cost-efficient than RL, even when the courses have been only partially converted to DL.[4] However, given that the magnitude of savings depends on the design and particular circumstances of a given DL program, coupled with the relative inexperience of the Army in implementing TADLP,[5] we assumed that DL

---

[3]For the purposes of this analysis, we consider front-end costs of DL as sunk. The Army has already committed to DL. The current question does not deal with how much DL is costing to develop, but rather with where to place those development monies to address important resource issues in DL training.

[4]For example, see the many references to such studies in Chapters 13 and 17 of Greville Rumble, *The Costs and Economics of Open and Distance Learning*, London: Kogan Page Limited, 1997.

[5]The costs of DL in the Army need to be calculated based on Army experience. DL involves such a complex tradeoff among cost, quality, program characteristics, and student characteristics that analysts have concluded that the financial outcomes of any economic study of DL cannot be directly applied to any other effort. See Rumble, pp. 151–152.

and RL cost the same on a daily basis. While this is a conservative assessment of what DL can achieve, it keeps us from moving toward solutions that depend on unproven savings or on the transference of training costs to students and units.

Note that our assumption on cost—that RL and DL cost per day is the same—does leave room for cost savings related to the reduction of course length. In fact, we assume in this report that DL reclassification courses in the Army can be 30 percent shorter than their corresponding AIT counterparts. The 30 percent number originated from research conducted by Orlansky and String, who examined the results of some 30 studies of the effects of DL on military training.[6] The Army Science Board, in its 1997 study of Army DL, also concluded that course length could be reduced 30 percent.[7] Finally, we tested the 30 percent assumption against early course designs in the TADLP, finding an average reduction of close to 30 percent. For further discussion of these points, see Appendix B.[8]

In practice, increases in the efficiency of learning could translate into superior performance rather than reduced course length. Many studies of DL have pointed out the tradeoff between training effectiveness and training time.[9] As DL develops in the Army, training managers and Army leadership are going to be presented with a large number of choices regarding the time versus the quality of training. For the purposes of this study, we need to balance our assumptions about training cost with corresponding assumptions about training quality. We address these assumptions below, along with a more general discussion of the DL quality issue.

---

[6]See Orlansky and String (1979).

[7]Army Science Board, "Distance Learning," Final Report of 1997 Summer Study, December 1997, p. 22.

[8]An indirect effect of the reduction in course length is an increase in the amount of manpower that is available to operational units and a reduction in the TTHS account. These potential savings are discussed in the companion research document, Leonard et al. (2001).

[9]For example, see Rumble (1997), Orlansky and String (1979, 1981), and Orlansky (1983).

DEALING WITH THE QUALITY DIMENSION OF DL

The analyses in this report (and in its companion report) carry with them the assumption that DL's potential can be fully realized in many of the Army's training programs without causing any reduction in the quality of training. Making this assumption allows us to fully explore how much DL might reduce shortages in Army occupations.

Moreover, we note that there is a considerable volume of past research supporting the contention that DL, if properly implemented, can provide training as effectively as the classroom training it replaces. For example, Barry and Runyan (1995) examined 11 studies of military courses and concluded that there was no significant difference between the performance of students in DL and RL versions of the same course. To cite another example, Phelps et al. (1992, pp. 113–125) found that knowledge gained in engineering and leadership courses offered to a group of Reserve Component officers was at worst not significantly different between RL and DL groups. Along the same lines, in a test of distance versus resident education on selected subjects from the Army Command and General Staff Officers' Course, Keene and Cary (1992, pp. 97–103) found that "students who received the distance learning instruction evinced superior knowledge of the subject matter at the end of the instruction." Finally, Farris et al. (1993) found that computer-based training could be used effectively in teaching many of the skills required for artillery fire direction specialists.

However, there can be no doubt that DL's introduction will bring about large and fundamental changes—technically, organizationally, and culturally—in how training (particularly but not only institutional training) is conducted. Viewed in this way, DL clearly poses some risks to the quality of training, especially during the transition period. The key to maintaining training quality and achieving desired learning outcomes will continue to be careful implementation and monitoring, which in turn will require continued emphasis and support throughout the Army. In the next paragraphs, we discuss areas where continued attention will be needed to uphold the overall quality of training and education as DL programs are introduced.

Maintaining the quality of training will require special emphasis on DL courseware in general. The shift to a greater dependence on technology (rather than instructors) to deliver training means a greater and more central role for courseware development and maintenance in maintaining training quality. Further, the fast development rate of new learning technologies effectively shortens the cycle time between needed course revisions. Finally, the role of courseware in the Army context becomes even greater considering the need for not only DL courseware, but also modularized, "just-in-time" training. In sum, if inadequate emphasis is placed on DL courseware, the quality of the training will be adversely affected.

DL initiatives must be implemented with due concern for retaining the benefits of residential learning where appropriate, and with careful selectivity in determining which portions of a given training program should be taught using DL. For example, many of the Army's RL courses, especially professional development courses, have important group process–oriented collaborative requirements. Losing these components in a DL-supported course could lead to a decrease in training quality. Some collaboration and group interaction can be built into DL segments of these courses, and consultations with instructors need not always be face-to-face (they aren't always in RL environments, either). But interactions over electronic media cannot fully substitute in every case for the value of direct face-to-face contact.

Judgments about DL conversion must also take into account some of the more intangible, but nevertheless real, benefits that RL conveys by allowing soldiers to associate in an academic environment with their peers and with subject matter experts. For example, reducing the length of residential training in some courses will reduce the opportunity for the Army's developing leaders to network with one another. While networking does not contribute directly to training quality per se, it does enable development of trust and confidence among peers that can enhance their effectiveness in their subsequent careers when they may be called on again to work together. This can legitimately be considered a contributor to the value of institutional training. The key to maintaining the overall effectiveness of the training program, then, will be to retain those aspects of direct interaction that cannot be replaced and to utilize fully the potential of new distance learning technologies to enable quality collaboration and interaction where needed. Application of this principle means there will be clear limits to the degree of DL conversion that would be appropriate.

Another key element of training effectiveness under DL will be to adequately redefine and support the roles of the student, the local commander, supporting installation activities, and the proponent schools. By moving more instruction out of directly supervised residential training environments and into the field, DL increases the responsibility of soldiers and their chain of command for ensuring that training standards are met in a timely manner (even though the schools will still play an important role even during DL phases). Adequately defining and supporting these new roles will be critical to DL success. Moreover, DL will create a need for other new or modified forms of support, e.g., "fenced" study time for students at home station, e-mail or Web-based academic aid and supplemental tutorial materials, periodic feedback for students, instructor help lines, and control of performance testing materials. Finally, while it may be possible to operate DL phases of courses with somewhat less administrative support overall, developing and supporting new administrative processes in the DL environment (e.g., for scheduling, enrollment, record keeping, certification) will also be critical to maintaining training quality. Failure to provide adequately for these types of support in DL-supported courses can lead to higher course attrition, longer completion times, insufficient learning or retention of important material, and, ultimately, lower training quality.

Finally, maintaining training quality under DL requires proper attention to incentive structures. As with any major innovation, the changes required to convert to DL will encounter natural resistance from some of the organizations and individuals entrenched in the current training paradigms. Resistance, in turn, can lead to a decrease in DL training quality. Providing the suppliers and users of training with financial and other incentives can help maintain quality by stimulating DL participation, fostering promising DL initiatives, and encouraging experimentation with new DL applications as they emerge.

In summary, while we hold in this report that DL can maintain high training quality, we also recognize that replacing resident learning with distance learning, if accomplished improperly, can lead to a reduction in training quality. In particular, we note the importance of avoiding the following implementation traps:

- Choosing inappropriate course segments for conversion.

- Developing courseware with inappropriate or outdated instructional media.

- Failing to make sufficient changes to existing processes and support activities to support DL's requirements.

- Failing to provide adequate resources.

- Providing insufficient incentives for students, commanders, and supporting activities to play their proper roles.

Chapter Four

HOW DL CAN IMPROVE THE EFFECTIVENESS OF RECLASSIFICATION TRAINING

As discussed in Chapter Three, reclassification is one of the strategies the Army employs to reduce personnel shortages and, in particular, to increase fill rates in shortage MOSs. It reutilizes existing manpower to fill force structure requirements. Reclassification courses give the Army an additional way to fine-tune its personnel distribution processes. Reclassification also gives soldiers additional flexibility in their choice of occupation, with the result that fewer valuable soldiers might leave the force.

In this chapter we first discuss the sources, process, and number of Army reclassifications and what benefits reclassification provides in reducing shortage MOSs. We then discuss how DL can help expedite the process and thus enhance the benefits of reclassification in this area. Finally, we discuss some of the potential forcewide benefits of DL reclassification.

SOURCES, NUMBER, AND PROCESS OF RECLASSIFICATIONS

Sources of Reclassifications

Reclassification candidates can come from a number of sources. One source is occupations in which the Army has a surplus. For example, although the force as a whole is short in SL1 soldiers, 18 percent of the MOSs in FY99 actually had a surplus. That surplus numbered some 6,250 soldiers at the MOS level of detail, two-thirds the number of shortages at that level. Table 4.1 shows some MOSs (grouped by Combat Arms versus other occupations) with surpluses at SL1, and the amount of the surplus.

Of course, for a soldier within a surplus MOS to reclassify into a shortage MOS, he or she must demonstrate an aptitude for the new occupation. What it takes to qualify varies and usually has to do with scores on one or more parts of the Armed Forces Qualification Test (AFQT). Applying qualification screens reduces the pool of available soldiers somewhat, but in most cases, many soldiers can qualify for other occupations. As an example, Table 4.1 shows the percentage of soldiers in each surplus occupation that would qualify for reclassification into the 67T MOS, one of the critical shortage occupations and one of our selected case MOSs. In this case, qualification means scoring 105 or more on the motor maintenance section of the AFQT. Forcewide, we found that 55 percent of the existing force met that requirement.[1]   Among occupations with surpluses (shown in

Table 4.1

Surplus MOSs and Their Potential as a Source for Reclassification into 67T

MOS	Title	E3–4 Surplus	Percent Motor Maintenance Score > 105
Combat Arms occupations			
11B	Infantryman	1,412	62
19K	Armor Crewman	532	57
13B	Canon Crewmember	452	42
11C	Indirect Fire Infantryman	380	59
Other occupations			
91B	Medical Specialist	905	55
51M	Firefighter	92	59
51B	Carpentry and Masonry Specialist	45	47
14T	Patriot Missile Crewmember	47	54
71G	Patient Administration Specialist	33	37

SOURCES: FY99 shortages file and EMF.

---

[1]AFQT scores are kept in the Army's Enlisted Master File (EMF).

Table 4.1), 37–62 percent would qualify for service as a helicopter repairer.

A second source of reclassifications is soldiers from occupations that are phasing out of the Army. This type of reclassification was highly prevalent during the drawdown years in the early and mid-1990s, but it still occurs today in more modest numbers. For example, in the 67 Career Management Field (CMF), three occupations are phasing out because the equipment is leaving the active inventory: 67N, UH-1 Repairer; 67V, Observation/Scout Repairer; and 67Y, AH-1 Repairer.

In these cases, soldiers are encouraged (but not required) to reclassify into occupations with shortages, especially when the shortage occupations are functionally similar to the occupations phasing out. In the case of CMF 67, shortages exist in all helicopter repair occupations. As a result, soldiers from 67N, 67V, and 67Y can conveniently reclassify into one of the remaining helicopter repair occupations, like 67T.

A third source of potential reclassifications is soldiers who might otherwise leave the force because, though otherwise qualified, they are dissatisfied with their current job classification. The "reclassification option" at reenlistment provides soldiers some flexibility and freedom in changing their occupation. The choice is offered mainly to soldiers in surplus occupations willing to reclassify into shortage occupations. Reclassification from surplus to balanced occupations or from balanced occupations to balanced occupations is also sometimes allowed. Inasmuch as the reclassification option reduces job dissatisfaction as one potential source of a decision to leave the Army, the result is that fewer soldiers leave the force.

A final source of reclassifications, if one is willing to use the term "reclassification" in a broader sense, is soldiers who have already left the force but are considering reentry. Technically termed "prior-service accessions" rather than reclassifications, these soldiers are included here because their retraining requirements are similar to those of currently serving soldiers who reclassify (i.e., they have proven experience in the Army but require training in a new occupation) and because their retraining has a similar purpose (i.e., to fill MOS shortages).

The Number and Process of Reclassifications

The Army reclassifies a sizable number of soldiers each year. In FY99, the total number of reclassifications was 5,220 soldiers.[2] To provide some perspective about the importance of reclassification for filling shortages, without reclassification, the total number of Army shortages would have been 55 percent greater.[3] There were also an estimated 2,910 prior-service accessions trained to fill MOS shortages.[4] It is currently unknown how many more would reclassify or reaccess if the programs could be made more attractive to potential participants.

To reclassify, existing soldiers currently have to go back to Advanced Individual Training (AIT) courses, the same courses used to provide an initial skill to new recruits. Special reclassification courses for experienced soldiers, though long available to RC soldiers, have been virtually nonexistent for the AC. Similarly, on-the-job training (OJT) is not available, since it was discontinued as a training option for reclassification in the mid-1980s. Considering both reclassifications and prior-service accessions, nearly 10 percent of AC soldiers in AIT courses are actually experienced soldiers rather than new recruits.[5] These soldiers must complete the full AIT course even though, for them, portions are redundant.

Figure 4.1 shows the time-in-service point for the 5,220 reclassifications that occurred in FY99. The figure shows both the year of service and whether the reclassification occurred at reenlistment (the darker part of the columns) or during a term of service (the lighter part of the columns). Most of the reclassifications (3,250, or 62 percent) occur at reenlistment points (as explained above), usually during the third or fourth year of service. The other 38 percent of reclassifications occur during a term of service rather than during a reenlistment

---

[2]Reclassification Management Branch, PERSCOM.

[3]Instead of 9,400 shortages, the number would be 9,400 + 5,200 = 14,600, a 55 percent increase over 9,400. This is a maximum estimate; it assumes that no reclassified soldiers were assigned to balanced or surplus occupations.

[4]The source is an estimate (run 9901) from the MOS Level System (MOSLS), a dynamic inventory projection model of the enlisted force at the MOS level of detail.

[5]The source here is the number of AIT inputs taken from the Army Training Requirements and Resources System (ATRRS).

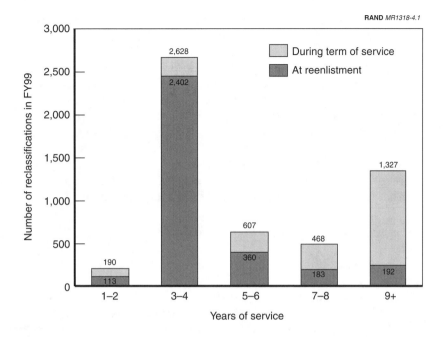

Figure 4.1—Soldiers Reclassify at Various Points in Their Careers

window. These tend to occur later in a career; in fact, about 20 percent of all reclassifications are for soldiers with nine or more years of service.

Not all mid-term reclassifications are related to or motivated by the need to fill MOS shortages. For example, some CMFs, like special forces (CMF 18), come into being primarily through reclassifications from other MOSs rather than through direct accession. In addition, some mid-term reclassifications are actually mandatory reclassifications that occur when soldiers cease to meet minimum requirements for a MOS (e.g., because of physical capability, security clearances, or licensing requirements).[6]

Nonetheless, about 10–20 percent of the mid-term reclassifications are voluntary reclassifications specifically undertaken to reduce MOS

---

[6]See DA PAM 611-21 for a fuller explanation of mandatory reclassification.

shortages. Most of these occur in the FAST TRACK program, where soldiers from surplus MOSs (or surplus grades within MOSs) are targeted and formally offered reclassification into shortage MOSs. One incentive for those offered reclassification is the better promotion possibilities in shortage MOSs relative to surplus MOSs. Experience in this program in recent years has shown that it takes four to five offers to get one soldier to accept the invitation. Said another way, 20 to 25 percent of soldiers invited to reclassify in FAST TRACK actually accept. When goals are not met with invitations (voluntary reclassifications in the FAST TRACK program generally redress only about 80 percent of shortages), mandatory reclassifications are ordered to further reduce the surplus and fill the shortages identified.

BENEFITS OF RECLASSIFICATION

The reclassification strategy has a number of benefits. First, for the purpose of reducing shortages, it is more productive for the Army to reclassify a soldier than bring in a new soldier through the accession process. Second, reclassification can be targeted to reduce shortages for SL1 and NCOs alike. Third, reclassification is more efficient than accession per SL1 shortage filled. We discuss each of these points in more detail below.

Reclassification Is More Productive Than Bringing in a New Soldier for Filling SL1 Shortages

To examine the effectiveness of increasing reclassification as a policy tool, we compared it to the approach of increasing accessions, assuming for the analysis that the reclassifications would occur, about equally, at the E3 and E4 level.

Figure 4.2 summarizes the results of the analysis, showing the steady-state effect on inventory of reclassifying 100 E3–4s per year and of bringing in 100 new accessions per year. (This figure illustrates 67Ts.)[7] First, note that both reclassification and accession are effective tools for reducing shortages. The effect of adding 100 sol-

---

[7]This example used data from MOS 67T. While inventory gains for different occupations will vary depending on the continuation and promotion rates of those groups, the total magnitude of this result would occur for almost any occupation.

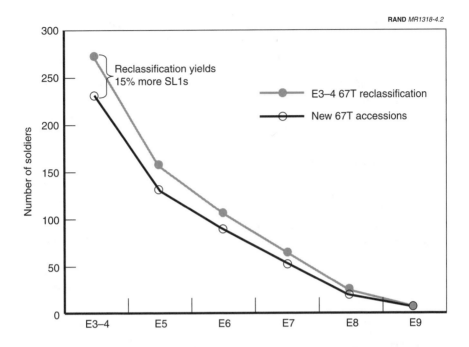

Figure 4.2—Long-Term Effect on Inventory of 100 Accessions per Year
Versus 100 Reclassifications

diers on SL1 (E3–4) alone is an eventual increase in inventory of 240–270 soldiers (the two leftmost dots in Figure 4.2). Over all grades and skill levels (that is, adding up all the dots), the effect of accessing or reclassifying 100 soldiers per year increases inventory in the steady state by 5 or 6 times that number.

Comparing accession with reclassification, Figure 4.2 shows that the Army would get somewhat more returns (about 15 percent) from the reclassification strategy than from the accession strategy. The primary reason is that the reclassification strategy avoids nearly all the losses during initial entry training that occur in the process of accessing new soldiers into the Army. Soldiers undergoing reclassification have proven ability and affinity for Army life, which new recruits have yet to demonstrate.

Reclassification Also Works for Reducing NCO Shortages

Having examined shortages in junior personnel, we now turn to the consideration of NCO shortages. How much could reclassification alleviate shortages at the NCO level? Figure 4.3 sheds some light on this question. In this case, we compare the steady-state effect on inventory of 100 E5 reclassifications per year versus 100 E3–4 reclassifications per year (The E3–4 line is a repeat of the line shown in Figure 4.2). Note first that reclassifying E3s and E4s does, by itself, eventually lead to a larger number of NCOs. In fact, adding the amounts for E5 and above in the E3–4 line, we find that more than half the eventual gains (56 percent) of E3–4 reclassification are at the NCO level versus 44 percent at the SL1 level. This occurs because many of the reclassified soldiers stay in the force long after their transition to a new occupation.

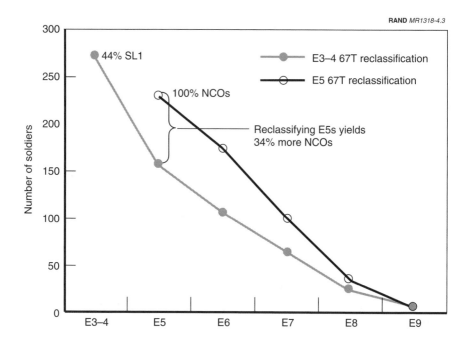

Figure 4.3—Long-Term Effect on NCO Inventory of 100 E3–4
Reclassifications per Year Versus 100 E5 Reclassifications

However, obtaining new NCOs from reclassifying at the E3–4 level depends on "growing" the newly reclassified junior personnel, a process that takes a number of years to complete. When occupational shortages are at the NCO level, a strategy that deals with NCOs in the near term is most immediately effective and therefore may be more appropriate. To deal with such situations, the Army has allowed for soldiers in many occupations to reclassify at promotion to E5. Reclassifying E5s (or higher grades) allows *all* the inventory increases to occur at the NCO level (as shown by the black line in Figure 4.3), and thus some NCO shortages can be filled immediately. Moreover, in the long term, reclassifying E5s yields 34 percent more NCOs than the E3–4 reclassification case.[8]

Reclassification Is More Efficient Than Accession

In addition to its greater effect on inventory, the reclassification strategy costs less per position filled than the accession strategy (for SL1 shortages) or the bonus strategy (for NCO shortages). The savings are greatest in cases of voluntary reclassifications unconnected with reenlistment decisions. In such cases, shortages are filled within the continuing force, without having to draw on the pool of newly accessed or reenlisted soldiers. In effect, such reclassifications make more efficient use of a given force or, viewed another way, reduce the cost of force structure imbalances that inevitably occur in a dynamic personnel distribution system. One way to estimate the value of such reclassifications is to use the cost avoided by not using additional soldiers. For example, for an E4 with 3–4 years of experience, the annual cost avoided (for each year the soldier fills the shortage position) is nearly $32,000.[9]

In all cases of reclassification, the cost of producing an additional qualified soldier is less than that for accession or reenlistment using

---

[8]This argument is not meant to imply that obtaining more NCOs by reclassifying at E5 is preferable to reclassifying at E4. While reclassifying at E5 yields more NCOs in a shorter period, reclassifying at E4 leads to more qualified NCOs due to their longer experience in the occupation. Decisionmakers must still consider the tradeoff in deciding how to obtain more personnel.

[9]Regular Military Compensation (RMC) from the 1999 Military Services Almanac (which includes basic pay, housing and subsistence costs, and the value of military tax advantages) and an estimate of the cost of retirement accrual.

a bonus incentive. The argument for the former case is illustrated in Figure 4.4, which documents the lesser cost of reclassification com-pared to accession in the case of 67T. In both cases, the cost of skill training is the same: $22,000 per trained 67T soldier. These costs include those for military and civilian training personnel (for in-struction and training support) and those for operations and main-tenance (O&M), ammunition, and Base Operating Support (BOS).[10] However, in the case of accessions (the left bar), costs include en-listment bonuses and the cost of basic training, costs that the re-classification strategy avoids (note their absence in the right bar). In addition, student training pay is higher in the case of accession because of the additional time required for basic training.[11]

Comparing the costs of reclassification with reenlistment through SRBs is more complex than comparing the costs of reclassification with accession. First, NCOs who are retained in their original occu-pation presumably have greater capabilities (at least initially) than NCOs who reclassify to that occupation. This is not necessarily true when comparing newly accessed versus reclassified soldiers at skill level 1. Second, calculating the cost of bonuses is complicated by the difficulty of targeting individuals who would stay in the Army due to the existence of the bonus. Because soldiers' reenlistment intentions cannot be known beforehand, bonuses are normally paid to all sol-diers within an occupation and grade, some of whom would have stayed in the Army even without the bonus. Thus, the cost of retain-ing one more soldier is the cost of his or her bonus, plus the cost of the bonuses paid to soldiers who would have stayed anyway.

Making a gross estimate of the average cost of bonuses in the case of 67T, we conclude that reclassification is less costly than bonuses as a way to fill shortages. We used the average bonus for soldiers in Zone A, multiplier 1.5 (the multiplier in effect at that time for 67T), which

---

[10]That amounts to about $1,500 per week for a 15-week AIT course.

[11]The cost of accession would be several percentage points greater if training attrition "overhead" costs were considered. In the case of reclassification, almost all soldiers who enroll in the reclassification course graduate from that course. But in the case of accession, soldiers have to pass through the initial "filter" of basic training, where attrition is significant. In addition, more first-time soldiers taking AIT fail to graduate than do soldiers reclassifying. Thus, the training attrition cost of obtaining an SL1 soldier is higher for accession than for reclassification. However, for simplicity, attrition costs were not considered in Figure 4.4.

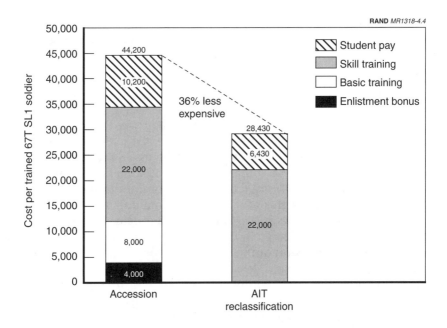

Figure 4.4—Reclassification Has Less Acquisition and Training Cost Than Accession to Alleviate SL1 Shortages

was $7,200 in FY99.[12] To address the issue discussed above, we assumed that retaining one more soldier required giving bonuses to that soldier plus five others; that is, five of six soldiers would have reenlisted without bonuses.[13] Under this assumption, the average cost of filling one shortage position using bonuses is $43,200 ($7,200 × 6), about 30 percent more than the cost of training ($22,000) and

---

[12]The source is the Retention Management Division, Enlisted Personnel Management Directorate in PERSCOM.

[13]This assumption is derived from data received from the Retention Management Division, Enlisted Personnel Management Directorate in PERSCOM. We were told that the estimates of the effects of bonuses on retention had wide confidence intervals, since they are based on a study done in the early 1980s. To avoid overestimating the cost of bonuses, we chose a figure at the high end of the interval, which implied bonuses were extremely effective at increasing retention. If bonuses are less effective, then the estimated cost of using bonuses to fill shortages is higher than estimated here.

military pay and allowances ($11,000 for an E5 with 6 years of experience) for a soldier who reclassifies in the 15-week 67T AIT course.

HOW DL CAN ENHANCE THE RECLASSIFICATION STRATEGY

The analysis above shows why the Army already uses the reclassification strategy. The focus here is on whether DL can enhance the benefits of the reclassification strategy. If DL could make reclassification more attractive to students and commanders, the strategy would (a) be able to redress a larger portion of the shortages and (b) be able to accomplish this at a lower incremental cost than exists today. In this section we argue that DL could provide a more attractive reclassification strategy in three ways. First, we argue that DL can lead to an expansion of the reclassification program because it can offer a shorter course, at least some of which could be conducted closer to home (e.g., at the home unit, at a Total Army School System (TASS) battalion, etc.). Second, DL could alleviate equipment bottlenecks at AIT because of the potential availability of unit equipment for training. Third, DL reclassification will make the process of filling shortages more efficient overall, both from the personnel acquisition and training standpoint and because it may lead to a better utilization of bonus funds. The first two points are described in more detail in the first subsection below; the final point is discussed at length following that.

DL Courses Could Expand the Reclassification Program and Reduce AIT Equipment Bottlenecks

In general, DL reclassification courses can be shorter and more flexible than their RL counterparts, and they can involve less time away from home station. This lowers the transaction cost (in terms of time, dollar cost, and inconvenience) of filling a shortage, both for potential reclassifiers and for their commanders. Basic economics argues that lowering the transaction costs involved in obtaining a product will increase the amount demanded. Thus, the introduction of DL reclassification courses could allow the expansion of that method of filling MOS shortages.

Table 4.2

DL (TADLP) Versus AIT Course Characteristics: The Example of the 67T Reclassification Course

Characteristic	AIT Course	DL Course (TADLP)
Total course length	15 weeks	8 weeks, 3 days
Residential length	15 weeks	4 weeks, 1 day
DL length	None	4 weeks, 2 days
Testing out of already mastered material	No	Potential Yes
Potential obstacles	• Funding • Training seats • Equipment	Cost of added course development

SOURCE: TADLP IPR November 1999.

Table 4.2 shows an example (67T) of a forthcoming DL reclassification course compared to the AIT course in the same occupation. Characteristics of the DL course were reported during the October 1999 DL In-Process Review (IPR) at TRADOC. The DL-supported course is 43 percent shorter in total (8 weeks, 3 days versus 15 weeks). Moreover, the DL course is 72 percent shorter in terms of time away from home station—only 4 weeks, 1 day.

Further, with a modularized DL course, there is potential for DL to shorten the course even further in cases where the trainees have already mastered some of the required skills. This could occur for 67N, 67V, and 67Y personnel, for example, who are already skilled helicopter repairmen but are reclassifying into new helicopter repair MOSs as the equipment mix of the force changes.

Finally, a DL course could overcome some of the current obstacles to undertaking more reclassification through the AIT course.[14] First, a

_____

[14]Of course, cost issues can arise on the DL side as well. For example, although we are assuming that from the perspective of the DCSPER the front-end investment costs of DL are sunk (see the discussion in Chapter Three), additional development costs might accrue if the DL program turned out to require more funding than currently allocated to address DCSPER concerns about DL design.

lack of training seats is less likely to be an issue because students spend less time in residence. Similarly, the current equipment shortage at Fort Eustis (the AIT base) is less likely to be an issue if some of the hands-on training could be completed at the soldier's unit using unit equipment. Finally, a shorter DL course is less likely to have problems with funding. Currently, a lack of available funds can prevent reclassification training (and cross-training) in cases of "TDY and return" (as opposed to training between assignments). For TDY and return, the training funding must come from the installation and compete with its other priorities.

DL Training Will Make the Process of Reducing Shortages More Efficient

As mentioned above, DL training offers two different ways of making reclassification more efficient at filling shortages: (1) by being less expensive from an acquisition and training perspective and (2) by potentially saving on bonuses needed for reenlistment. Each savings is discussed below in the context of SL1 reclassification.

DL reclassification training is a less expensive way to reduce shortages. First, if a more attractive DL course can lead to more voluntary reclassifications among the continuing force, force efficiency can be increased and the cost of force structure imbalances reduced. As argued above, in this situation the number of shortages can be reduced without the Army having to draw on newly accessed or newly retained soldiers.

Second, DL reclassification can further reduce the cost of filling shortages. Figure 4.5 makes this point, expanding on the graphic used earlier in Figure 4.4 to illustrate the lower cost of DL reclassification compared to accession and traditional reclassification in the case of 67T. While the cost of skill training is the same in the cases of accession and AIT reclassification, it is more than 43 percent less for DL reclassification because of the shorter course length. As a result, DL reclassification costs 43 percent less than AIT reclassification, and 64 percent less than the full cost of accession. Because median course length reductions across all courses are expected to be closer to 30 percent (rather than the 43 percent observed in the case of

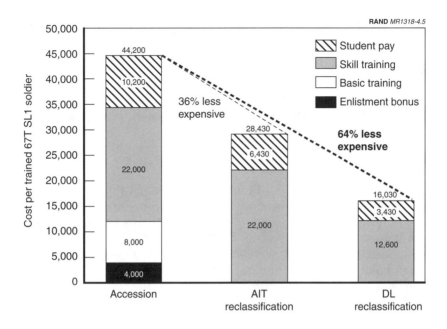

Figure 4.5—DL Reclassification Is the Least Expensive Way to Alleviate SL1 Shortages

67T), the proportional cost reduction in other occupations will be less, but still substantial. [15]

DL reclassification might lead to a more effective utilization of reenlistment bonuses. Another benefit from using DL to increase the number of reclassifications is a more effective utilization of SRBs. We are speaking here about the potential to reallocate SRBs to fill even more shortages. The reallocation becomes possible due to an indirect effect of SL1 reclassifications in selected occupations. Specifically, this could occur in cases, such as 67T, where a MOS has an SL1 shortage (inviting increased reclassifications) but avoids an NCO shortage through the use of SRBs to boost retention. In such cases, the eventual movement of the additional SL1 reclassifications to NCO positions could lead to a local surplus of NCOs. The exis-

---

[15]See the discussions in Chapter Three and Appendix B on estimating the cost of DL versus RL courses.

tence of the surplus, in turn, would allow reallocating SRBs to reduce shortages in other occupations.

Figure 4.6 estimates the effect of eliminating the current 1.5A bonus for 67T after increasing reclassifications. The main point is that reclassifying soldiers (or cross-training soldiers or consolidating MOSs) to overcome the SL1 shortage has the additional advantage of producing an adequate number of NCO leaders in this occupation without the aid of a bonus. This frees up bonus monies to be applied to other occupations.

The white columns in the figure show the 67T original inventory, compared to the line, which represents authorizations. The gray columns show the steady-state effect of 95 reclassification training graduates per year, the number needed to eliminate the SL1 shortage (note that the gray column reaches the line for E3–4s). Finally, the black columns represent the estimated inventories after eliminating the level A bonus.

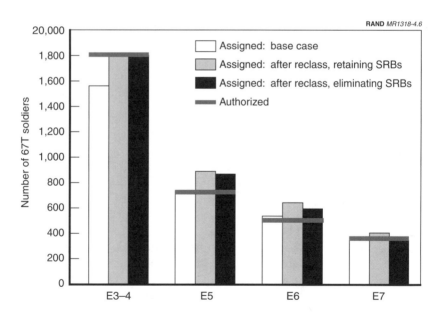

Figure 4.6—More SL1 Reclassifications Through DL Can Eventually Lead to NCO Surpluses, Reducing the Need for NCO SRBs for 67T

The difference between the gray and black columns in the figure is 125 soldiers (in E5 through E7 combined), the estimated number of 67Ts who would separate from the Army because of the elimination of the bonus for NCOs.[16] However, note that despite the loss of NCOs, the final number is still either above or at the authorized inventory levels, as shown by the relationship of the black column to the line. In other words, eliminating the bonuses to NCOs in 67T cuts the surplus created by the SL1 reclassifications, but does not eliminate it.

Figure 4.7 shows, based on the information in Figure 4.6, the potential amount of bonus dollars that might be reallocated from 67T to other occupations with shortages. The white part of the column on the left shows the 125-soldier reduction in NCO inventory—the steady-state, net effect of the bonus elimination illustrated in the last figure. The black part of the column estimates the number of NCOs in the steady state electing to stay in the force who would no longer receive the bonus during their first reenlistment period. (It is based on the assumption of a four-year initial reenlistment). At the far end of the figure we see the yearly bonus dollars made available for reallocation, $1.47 million, calculated by multiplying the number of soldiers in the parts of the bar by 1.5 (the bonus multiplier in effect in October 1999) times their monthly basic pay.

POTENTIAL FORCEWIDE BENEFITS OF DL RECLASSIFICATION

DL reclassification can increase readiness by providing a better vehicle for reducing the number of personnel shortages. It can also increase the efficiency of the overall process for filling shortages.[17]

---

[16]The 125 estimate is an upper-bound estimate derived from data received from the Retention Management Division, Enlisted Personnel Management Directorate in PERSCOM. However, since the estimate is based on a study completed in the early 1980s, it must be considered an extremely rough approximation. The exact effect that the reduction or elimination of bonuses in certain MOSs would have is a matter for further study and beyond the scope of this report. The point here is mainly that in cases where surpluses are produced by SL1 reclassifications, some reallocation of SRBs is likely to be possible.

[17]Other benefits of reclassification, like increasing the availability of soldiers to commanders and reducing the size of the TTHS, are analyzed in Leonard et al. (2001).

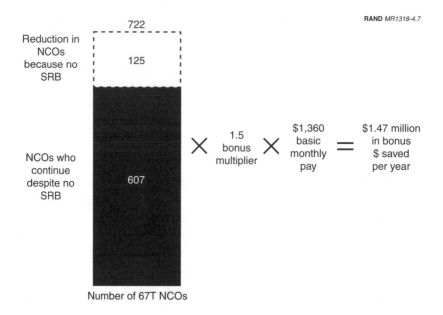

RAND *MR1318-4.7*

Figure 4.7—Eliminating the SRBs in 67T with DL Reclassification Would Make Funds Available to Reduce Other Shortages

The size of the total benefits of DL reclassification can be substantial, but the benefits are of varying types, each with important characteristics and distinctions. Here we summarize those benefits by type and discuss their potential impact forcewide.

Reduction in the Number of MOS Shortages

First, our analysis suggests that DL could stimulate an expansion of the Army's reclassification program, resulting in a corresponding reduction in the number of MOS shortages. The basis for this expectation boils down to a transaction cost argument. For potential trainees, the transaction costs for entering a new occupation are reduced because DL courses take less time to complete and involve less time away from home and family than the traditional AIT course. For commanders, the transaction costs to retrain surplus unit personnel (to increase unit personnel readiness) are similarly decreased because the training costs less, takes less time to complete, and allows some access to the soldier during the training period. Basic economics argues that lowering the transaction costs of employing the strategy will increase the amount demanded.

There are two ways that the number of reclassifications might increase to fill more MOS shortages. First, as argued above, the proportion of soldiers who accept reclassification when offered might go up. Currently, reclassification attracts relatively few. In the Army's current voluntary program (see the discussion of FAST TRACK earlier in this chapter), only about 20 to 25 percent of the soldiers asked accept the offer to reclassify. Moreover, only a small percentage of soldiers choose the reclassification option at reenlistment.

A second way to increase the number of reclassifications is to expand the window of opportunity for reclassification. For example, most reclassifications currently occur only after several years in the force; only 3.6 percent (190 out of 5,220) occurred in the first two years of service in FY99. The existence of the DL option for reclassification may make earlier reclassifications (ones that occur nearer the beginning of the first soldier assignment) attractive to soldiers or their commanders as a way to reduce shortages.

Estimaing how much DL might affect individual decisionmaking (those currently in the force in surplus occupations) would require some experimentation. However, we do know that benefits of additional reclassification on a per-capita basis can be substantial, and may even justify offering a reclassification bonus. For example, reducing force structure imbalances with more voluntary reclassification might be valued at $32,000 per soldier per year (the pay and allowances of an E4 with 3–4 years of experience who moves from a surplus to a shortage position).[18] That amounts to a yearly million-dollar savings for each additional 31 voluntary reclassifications.

Reduction in Cost of Reclassification

DL reclassification courses could save training resource manpower and dollars on the additional training load it would stimulate. Our

---

[18]While there are many other ways to try to quantify the value in readiness of filling one more shortage, we chose the cost of the soldier filling the shortage as the basis for this report.

analysis suggests that DL could reduce the cost of training for reclassification by 30 percent. This assumes that, on average, RL and DL training cost the same on a daily basis, but DL courses can be 30 percent shorter than RL courses (see the discussion in Chapter Three and Appendix B). To estimate the costs avoided on a per-capita basis, we assumed an average 10-week AIT course, at $1,500 per week,[19] reduced to a 7-week DL reclassification course. In that case, $4,500 in training costs would be avoided for each additional soldier reclassified. That amounts to a million-dollar savings for each additional 222 reclassifications.

In addition to reducing the cost of training additional reclassifications, DL could save on the cost of training those who currently reclassify. To estimate how much, consider FY99 as a base case. During that year about 3,550 soldiers reclassified[20] and 2,910 prior-service soldiers rejoined the Army. If we assume nearly all these soldiers could be trained with DL reclassification courses rather than AIT courses (i.e., full DL implementation), and that the same $4,500 in training costs per student could be saved (as per the argument above), training via DL reclassification would cost about $29 million per year less than training via AIT ((3,550 + 2,910) * $4,500).[21]

Although the savings calculated above reflect the potential benefits of DL in handling the current training load, they do not necessarily translate into additional budget or TDA manpower savings. The extent of actual savings depends on assumptions in the current budget. For example, if training resources have already been reduced in anticipation of DL and built into budget and manpower estimates, then the advantage of implementing a DL reclassification program is not additional savings, but rather a way to make the training more feasible under the current resource plan.

------

[19]Derived from FY99 Military Manpower Training Report, supplemented by the annex on resource trends.

[20]This number is a subset of the total 5,200 reclassifications, excluding special service school actions (e.g., to create special forces, CMF 18).

[21]Most of this amount will take the form of the pay and allowances of military and civilian manpower.

Increase the Effectiveness of the SRB program

A third type of benefit of expanding reclassification through DL is the potential opportunity to reallocate SRBs to increase their effectiveness. As explained above, the opportunity for reallocation is made possible by MOSs with special characteristics. In particular, we seek MOSs that have an SL1 shortage (inviting increased reclassifications), but no shortage of NCOs (possibly because of the use of SRBs). We estimate that 5 percent of the 13,500 soldiers receiving bonuses in FY99 were in occupations that were short at SL1 but balanced or surplus at NCO levels. Assuming those NCOs received the average bonus, $6,700,[22] the potential for SRB reallocation due to DL could amount to as much as $4.5 million per year. These funds could then presumably be used to further decrease MOS shortages in the remaining occupations.

Bonus savings, like the training resource savings mentioned above, presume that the "right" courses are converted to DL. Moreover, the savings are even "longer term" than the training resource savings mentioned above, because achieving them also depends on "growing" reclassified junior personnel into NCOs. Finally, the amount of the savings must be considered as a rough estimate, since actual outcomes will depend on the choices of individual soldiers in the face of bonus changes, a subject whose analysis is beyond the scope of this report.

Finally, it is worth emphasizing that realizing any of these benefits requires creating DL reclassification courses that are sufficiently attractive to alter the decisions of soldiers (e.g., whether to reclassify, stay in the force, or reenter the force from a civilian status) and, to some extent, their commanders (e.g., whether to seek soldier reclassification or to send soldiers to DL reclassification courses). Some of the attractiveness is built into the characteristics of DL (shorter courses, more completed at home station, more flexibility in the timing of training and the availability of soldiers to commanders), but other important components will have to be built into the implementation process. For example, success in implementing DL will depend on high-quality courseware, adequate student support

---

[22]Information on bonuses received from the Retention Management Division, Enlisted Personnel Management Directorate in PERSCOM.

for taking on the additional responsibility of DL training, added support from commanders and units in hosting the training, and sufficient administrative support in scheduling courses, matching students to courses, and reporting training results.

Chapter Five

HOW DL CAN IMPROVE THE EFFECTIVENESS OF CROSS-TRAINING AND MOS CONSOLIDATION

When we looked in Chapter Three at the generic strategies that the Army uses to reduce personnel shortages, we identified cross-training and consolidation of MOSs as two options. Since both strategies attempt to produce more effective soldiers by making them more capable of performing a broader range of activities, we treat them here as one strategy for discussion purposes. With cross-training, soldiers already proficient in one MOS are trained to perform related activities in another MOS, so that they can informally fill in for that other occupation when necessary. With MOS consolidation, MOSs that perform similar activities are formally combined into one occupation, and soldiers in each of the old MOSs are given additional training to become proficient in all aspects of the new MOS.

In this chapter we follow a similar approach to the one we used in discussing the reclassification strategy. We begin by discussing the cross-training/MOS consolidation process; we then turn to arguing for the feasibility of the two and for their usefulness in reducing personnel shortages. Then, we discuss how DL can enhance the effectiveness of the strategies. Finally, we discuss some potential force-wide benefits of using DL for cross-training and MOS consolidation.

THE PROCESS OF CROSS-TRAINING AND MOS CONSOLIDATION

Cross-training and MOS consolidation work best in occupations with high overlapping functionality. In the 67T case, for example, there are six other occupations dealing with helicopter repairs. Consolidation or cross-training is clearly possible here. This is reinforced by

49

looking at the civilian workforce, some of whom work on Army helicopters, which already uses generic helicopter repairmen. These
civilian helicopter repairers are given access to special add-on training, as and when required, for specific equipment and assignments.

One key consideration in using a cross-training strategy is the organizational context. For example, it makes no sense to cross-train two
occupations unless soldiers in the two occupations are going to work
side by side. In the case of Army aviation this test is met, since future
plans call for multiple aircraft types in the same unit, a balanced
mixture of attack, reconnaissance, and lift helicopters.[1]

A necessary ingredient for consolidating MOSs is a workable training
strategy, which can only be accomplished if the occupations are
good candidates in the first place. For example, if occupations are
overconsolidated, the amount of material to cover can lead to development of SL1 courses that are impracticably long. Simply cutting
that course to make it more feasible is clearly not the solution, since
that would threaten the quality of the work performed and the readiness of units using those occupations. An answer in some cases is to
create generic training courses that are supplemented by specialty
training on selected types of equipment depending on soldier
assignments.

THE FEASIBILITY OF CROSS-TRAINING AND MOS CONSOLIDATION

Unlike reclassification, there is some concern that cross-training and
MOS consolidation may not be feasible strategies in and of them-
selves, since they have not been widely employed in the recent past.
Thus, we briefly discuss our view that the strategies can indeed work.

Consolidation of MOSs was much more prevalent during the recent
drawdown period and before, as evidenced by a reduction of the
total number of MOSs by 40 percent from 1986 to the mid-1990s.[2]

---

[1]See Maj. Gen. Anthony Jones, "Aviation Modernization Strategy—2000 and Beyond,"
Army Aviation, May 31, 2000.

[2]BG Adair, CSA IPR of the Army Development System (ADS) XXI Task Force, briefing,
May 2000.

Not all these reductions involved consolidations; some reductions simply reflected (as does the anticipated disappearance of 67V, 67N, and 67Y mentioned earlier) equipment leaving the Army. However, many did involve consolidation of two or more MOSs into a single MOS.[3]

Present-day evidence comes from the Army Development System (ADS) XXI Task Force. As of early summer 2000, proponents working on that task force had suggested the merger of 44 MOSs into 22.[4] These 44 MOSs cover 17 percent of current authorizations in the Army.

Finally, there is also evidence that cross-training and MOS consolidation are indeed feasible, even for very tough cases. An earlier RAND study (Wild and Orvis, 1993) examined the feasibility of using "field-based cross-training" (FCBT), which referred to two programs: (1) consolidation of two or more MOSs by combining specialties, and (2) a shift from school to OJT, where AIT courses are shortened and reoriented to focus on general "core" skills and where a formal OJT program is instituted at the field unit to compensate for the reduction in schoolhouse training.

The RAND study focused on a tough case—CMF 67, helicopter maintenance—which includes the 67T MOSs we focus on in this study. This career field is tough because it involves advanced technology, has a significant peacetime training mission, and entails an especially high prevalence of safety issues.

The premises of that study were that MOS consolidation broadens enlistees' skills through cross-training and that broader skills can potentially lead to more flexible work allocation at the maintenance unit and more flexible assignment of personnel to field units. The study found that selective consolidations in the CMF offer the best way to maintain readiness while reducing training and personnel costs. The report also emphasized that the consolidations should

---

[3]For example, most current MOSs in CMF 31 are the product of at least one consolidation between 1986 and 1999. Specifically, 31G, 31K, and 31V combined to form today's 31U; 36L, 29NV8, and the old 31F combined to form today's 31F; 31M and 31D combined to form today's 31R; 31N, 29M, and 29V combined to form today's 31P; and 29Y and 31M7A combined to form today's 31S.

[4]See Adair briefing, op. cit.

focus on highly transferable skills and that they should be aug-
mented with policies that facilitate cross-training. Finally, it stressed
that training packages to support flexible unit assignment of similar
MOSs could be used to enhance cross-utilization without formal
consolidation.

Beyond the evidence of feasibility from the RAND study, there is also
evidence in the civilian sector. As mentioned above, civilian aviation
organizations currently train all primary helicopter repair skills in
one occupation.

THE EFFECTIVENESS OF CROSS-TRAINING AND MOS CONSOLIDATION

Not only are cross-training and MOS consolidation feasible, they
provide an effective strategy for alleviating the effects of personnel
shortages. In this case, however, the strategy does not accomplish
this task by *creating more soldiers to assign*, as is the case for reclassification; instead, it accomplishes the task by *increasing the skill base
of soldiers in existing assignments*, thus making soldiers more efficient. To understand this, think of two specialized MOSs, A and B.
Workers in MOS A might stand idle because of a temporary lack of
type A work, while type B jobs queue up because of a temporary high
demand for the work. Alternatively, workers in MOS A might have
work, but only on low-priority jobs, while high-priority type B jobs
again queue up. Clearly, when personnel are capable of both types
of work, such instances can be remedied.

Another way to think about the effect of cross-training and MOS consolidation is that they minimize the impact of personnel shortages.
In the scenario discussed above, the shortages do not get reduced by
cross-training and MOS consolidation; however, the ability to use
cross-trained personnel to help redress workload imbalances can
render the MOS shortages less damaging to readiness.

Of course, cross-training and MOS consolidation may also actually
reduce shortages. A consolidation of functionally similar Army
occupations could simplify the assignment process, allowing a
reduction in the 9,950 shortages that currently exist because of personnel surpluses in nonshortage occupations. Moreover, the

increased efficiency of soldiers could eventually allow a decrease in requirements (i.e., the need for the assignments in the first place).

Table 5.1 helps to drive home this point by exploring the potential for cross-training and MOS consolidation to reduce shortages in the MOSs of CMF 67. The table lists functionally similar 67 MOSs and indicates their present status in relation to SL1 shortages and surpluses and NCO shortages and surpluses.[5] Considering all helicopter repairers together, the gap between SL1 authorizations and assignments is approximately 11 percent (computed by dividing 419, the number of shortages, by 3,885, the number of slots authorized). This is less than the gap in 67T by itself (14.4 percent) because the gaps in other occupations are smaller. Thus, if cross-training or consolidation could reduce workload (or requirements) by about 11 percent, the current shortage in the entire series could be eliminated.

It is worth noting that while all MOSs but one have an SL1 shortage, NCOs (the right side of the table) show a slight surplus. Thus, if a

Table 5.1

The Potential of Cross-Training and MOS Consolidation:
The Example of CMF 67

MOS	Title	SL1			NCOs (E5–7)		
		Auth	Assgn	Diff	Auth	Assgn	Diff
67T	UH-60 Repairer	1,804	1,544	−260	1,554	1,612	58
67U	CH-47 Repairer	684	665	−19	834	879	45
67R	AH-64 Repairer	802	758	−44	670	671	1
67S	OH-58D Repairer	453	400	−53	547	510	−37
67N	UH-1 Repairer	109	68	−41	109	84	−25
67V	Observation/Scout Repairer	6	21	15	21	36	15
67Y	AH-1 Repairer	27	10	−17	54	35	−19
Total		3,885	3,466	−419 (11%)	3,789	3,827	38

SOURCE: MOS Data Sheet, PERSCOM.

---

[5]Note that the final three occupations in the table are being phased out of the active force.

consolidation strategy required some equipment-specific training for soldiers taking on new assignments, the surplus in NCOs suggests that there are enough senior personnel present to support this training.

HOW DL CAN ENABLE MORE CROSS-TRAINING AND MOS CONSOLIDATION

As was true with reclassification, using DL in conjunction with cross-training and consolidation could make the existing options more attractive, but in somewhat different ways for each. For cross-training into functionally similar MOSs, the modular aspect of DL training would allow avoidance of part of the reclassification course dealing with common tasks, reducing even further course length, training repetition, and TDY time. For MOS consolidation, the way DL would help depends on how the consolidation is accomplished. If two functionally similar MOSs are simply combined into one (perhaps because of technological change), the use of advanced learning technology might contribute to the development of a workable training strategy. But if the concept is to produce a generic specialist across two or more specialties (as is true for helicopter repairers in the civilian world), DL could provide the equipment-specific training they need for a specific assignment without leaving their home station.

In addition, DL makes it easier for soldiers to stay current in skills relevant to the requirements of their current job and then refresh quickly on a different set of (hopefully similar) skills in the same MOS when they are moved to a unit with different equipment.

It is important to note that DL by itself does not make any given MOS consolidation or cross-training a good idea, any more than it makes reclassification a good idea. However, in cases where consolidation does seem to be feasible and useful in reducing personnel shortages (like those cited above), DL can make consolidation strategies easier and less expensive to accomplish. Below we discuss some of these benefits in more detail.

DL Cross-Training Is Even More Appealing Than Reclassification to Soldiers/Unit Commanders

In the previous chapter we discussed how using DL courses to do reclassification made those courses more appealing. Doing the same thing with cross-training courses raises their appeal as well, as illustrated in Table 5.2, which modifies Table 4.2. In this case, the boldface in the table represents the key differences between the DL reclassification course and the DL cross-training course. Those differences center around the ability to modularize the existing DL course so that soldiers can test out of already mastered material when being cross-trained and, thus, not need to take an entire course. The ability to modularize courses through DL also drives the result, indicated in boldface, that the already shortened lengths for the first three entries can be made even shorter.

DL Cross-Training Can Reduce Costs in the Same Ways DL Reclassification Does

Figure 5.1, which builds on Figures 4.4 and 4.5 in the previous chapter, shows that DL cross-training promises to be even less expensive

Table 5.2

DL (TADLP) Versus AIT Course Characteristics:
The Example of the 67T Cross-Training Course

Characteristic	AIT Course	DL Course (TADLP)
Total course length	15 weeks	**Maximum of** 8 weeks, 3 days
Residential length	15 weeks	**Maximum of** 4 weeks, 1 day
DL length	None	**Maximum of** 4 weeks, 2 days
Testing out of already mastered material	No	Yes, if course modularized
Potential obstacles	Funding Training seats Equipment	Cost of course development

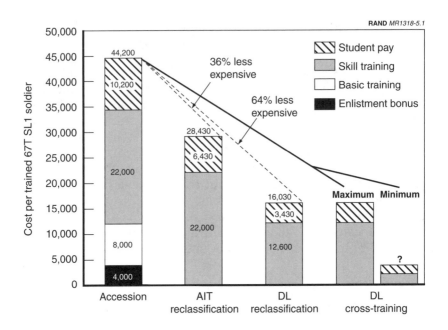

Figure 5.1—DL Cross-Training Is Even Less Expensive Than Reclassification for Alleviating SL1 Shortages

than reclassification for alleviating SL1 shortages. Just as we modified the right column in Table 5.2 with the word "maximum" to indicate that the courses are likely to be shorter in all key aspects, here we provide a range of maximum and minimum. The courses will be no longer than the reclassification course, in which case they would have the same cost levels; however, they could be quite a bit shorter, which would drive the actual costs per trained 67T SL1 soldier down much lower, depending on the degree of modularization. In addition, as was true for DL reclassification, DL cross-training can help make the SRB program more effective.

POTENTIAL FORCEWIDE BENEFITS OF DL CROSS-TRAINING OR MOS CONSOLIDATION

Using DL for cross-training and MOS consolidation can increase readiness by providing better vehicles to reduce the number of personnel shortages. As with reclassification, the availability of DL can make these strategies more attractive to consumers, and improve the efficiency of the overall process.

The primary benefit of using DL to implement these strategies is the avoidance of future training costs. How much can be saved depends on the extent these strategies are used across the force. We have already noted that the organization of future aviation brigades suggests compatibility with MOS cross-training and MOS consolidation. Similarly, we have noted that proponents working with the ADS XXI Task Force have submitted a list of 44 MOSs for consolidation, involving 17 percent of authorizations.

In addition, the ADS XXI Task Force will recommend future consolidation beyond the 44 already submitted. One of the goals of these consolidations is to make occupational positions easier to fill, reducing current shortages. Specifically, another 88 MOS consolidations are currently being considered or recommended for further study. The concept in these consolidations is to create a new MOS structure with bigger and fewer occupations facilitating more flexible personnel management policies. Positions within occupations will be filled with "adaptable soldiers" who have the ability to perform well in skills, knowledges, and abilities (SKAs) that span two or more functional areas. Each job within a MOS will not be considered separate; instead, it will be considered a refinement that can be modified by training through DL media (i.e., self-development or unit-conducted training) to bring the soldier up to required proficiency.

Clearly, the potential size of the benefits of DL cross-training and consolidation are large in the long run, once DL has been fully implemented. However, for benefits to be large during the long implementation period of TADLP will require coordination to ensure that the right courses are chosen for DL conversion and that the characteristics of the courseware fit the needs of the specific strategy employed. For example, we found that only 6 of the 44 courses initially proposed for consolidation by school proponents were on a course list for TADLP before FY03.

Chapter Six

HOW DL CAN ACCELERATE THE PACE OF PROFESSIONAL DEVELOPMENT TRAINING

When the Army reclassifies soldiers, it reduces personnel shortages by moving soldiers from surplus MOSs to shortage MOSs. When it cross-trains or consolidates soldiers, it helps alleviate personnel shortages by moving soldiers from nonshortage MOSs into similar shortage MOSs. However, there is another reason that some MOSs have shortage problems: Sometimes positions, while not vacant, lack fully qualified soldiers. We speak here about problems in the timing of NCOs completing the Army's BNCOC and ANCOC. If training for those who need it could be accelerated, then the shortages in trained personnel could be reduced.

In this chapter we look at the potential for DL to accelerate training, beginning by providing some evidence that there is a problem in this area and then showing how DL might help. We end, as before, by discussing some potential forcewide benefits.

ACCELERATION OF NCO TRAINING IS NEEDED TO REDUCE SHORTAGES

The size of the Army's E6 and E7 trained inventory would increase if soldiers needing BNCOC and ANCOC could be trained sooner. For FY99, we estimate that 8,500 E6 and E7 positions were occupied by soldiers not formally trained for those jobs or not trained for their grade. That number represents 2.4 percent of all authorizations, and 8.9 percent of E6 and E7 authorizations. For discussion purposes, we divide the number of untrained personnel into two groups. The first group comprises soldiers who have been promoted to grades E6 and E7 but have not completed the required BNCOC or ANCOC. The second group comprises soldiers in E5 and E6 grades who are serving in E6 and E7 positions, respectively, but have not completed the required BNCOC or ANCOC.

There is evidence on the size and characteristics of the second group in recent RAND work studying NCO leader development.[1] Figure 6.1 illustrates key results in those reports. RAND researchers examined the profile of NCOs assigned to operational units in November 1996. NCOs were divided into three groups—those serving above, at, or below grade (the grade the position requires is above, at, or below the grade held by the NCO)—and their promotion and separation rates were examined. Across E4s–E8s as of November 1996, we find, as shown in the figure, that about 10 percent of E4s–E6s indeed serve above grade and about half as many (in percentage terms) E7s do. Some, but not all, above-grade NCOs had already been selected for promotion to the next-higher grade and are eligible to attend formal NCO schooling.

There are a couple of reasons for this problem. One has to do with high personnel tempo (PERSTEMPO) demands. Given the impor- tance of senior leaders in units, it is often difficult for NCOs to get the time to go away for extended periods for training. A second reason is that some of those serving in positions for which they have not been trained are upwardly substituted "fast-trackers," mostly filling higher-grade positions for which there would otherwise be a short- age. The RAND research establishes the "fast-tracker" label by look- ing at one-year promotion rates for these three groups. For example, if we zero in on E6s who have between 7 and 10 years of service, we find that those serving above grade have a 16 percent one-year pro- motion rate, while those serving at grade have a 5 percent rate and those serving below grade have a 1 percent rate. Comparable E5 promotion rates are 20, 13, and 10 percent, respectively.[2]

---

[1]Research results are described in more detail in Shukiar, Winkler, and Peters (2000) and Winkler et al. (1998).

[2]Note that these are one-year promotion rates and not overall measures of promotion probability.

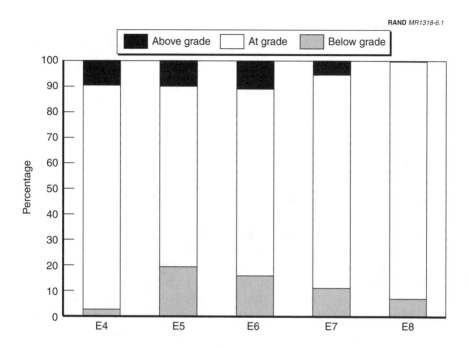

Figure 6.1—A Number of NCOs Serve in Positions Above Their Grade, Without Formal Training

Even more surprising, the one-year separation rates for the three groups are about the same for E6s and E5s—about 30 percent. In other words, those serving above, at, or below grade tend to leave at about the same rate. This raises the question of what the Army can do to ensure that it retains these fast-trackers at higher rates than those of the overall group. This is especially important for fast-trackers in hard-to-retain, shortage CMFs.

The current process for dealing with fast-trackers exacerbates the NCO shortage problem. Such fast-trackers are frequently not formally trained, depending instead on OJT until formal promotion and training can occur; unfortunately, the lack of timely training/promotion/compensation reduces the incentive for these soldiers to stay in the Army, which in turn leads to the disappointing retention rates noted above.

In addition to some NCOs serving in positions above their grade before training and promotion (the fast-tracker problem), other NCOs are serving in positions at their grade level but are not trained for their grade. While the Army would prefer to train all soldiers before they promote them, in practice, shortages in time and money keep it from reaching that goal.[3] Figure 6.2 illustrates this point and further supports the notion that NCO training needs acceleration. Only 76 percent of the E6s were trained before promotion in FY99. An additional 16 percent were trained within one year after promotion, and 1 percent were trained more than a year after promotion. Seven percent have no record of any training after their promotion, in either the personnel records (EMF) or the training records (ATRRS). Even fewer E7s were trained before promotion, 33 percent. Most (63 percent) were trained within the year after promotion. Three percent were trained more than one year after promotion, and 1 percent have no record of training completion. These figures further support the contention that NCO training needs acceleration, especially given the fact that the new NCO Educational System (NCOES) model is expected to require even more NCO individual training.[4]

DL CAN HELP ACCELERATE TRAINING

DL could make training possible earlier in the select-train-promote sequence. First, DL training can begin before scheduled residence training courses are available. Second, DL training can be taken in small pieces, on a "continuous" basis. Third, DL training can occur at home station. Fourth, modularized DL courses allow "testing out" of already mastered material, which means that fast-trackers who get much of their experience through OJT would not have to sit through the parts of course material they have already learned. Finally, DL can enhance self-development training. While self-development training is one of the Army's three pillars of leader development[5]

---

[3]Current Army policy gives E6s and E7s one year (and occasionally longer) from the point they are promoted to complete BNCOC or ANCOC.

[4]BG Adair, CSA IPR of the Army Development System (ADS) XXI Task Force, May 2000.

[5]The three pillars are institutional education, operational experience, and self-development.

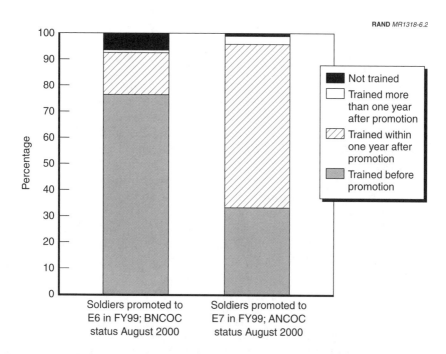

RAND *MR1318-6.2*

Figure 6.2—Many NCOs Are Schooled After Promotion

and is a promising way to train fast-trackers (who are somewhat more likely to be motivated toward self-development), it needs further refinement to be useful in this context, focusing more on military and leadership training.[6]

Using DL to accelerate NCO training will reduce shortages of trained personnel. First, DL can reduce the time to training completion, thus decreasing the number of untrained personnel. Second, DL training can increase the training readiness of fast-trackers. More specifically, it can support OJT but be monitored by the schoolhouse and, as mentioned above, could provide modularized training that allows "testing out" in areas where OJT has already occurred. And third, DL

---

[6]For further discussion, see the other report from this research project, Leonard et al. (2001).

training for BNCOC and ANCOC can set the stage for grade-specific reductions in shortages.

This last point is illustrated in Figure 6.3, which shows how earlier ANCOC training for the Microwave Systems Operator/Maintainer MOS (31P) could help enable the reduction of grade-level MOS shortages. The three columns in the figure show the number of assigned soldiers in the 31P MOS who are E5s, E6s, and E7s. The line shows how many positions are authorized for each grade, and the contrasting hatched pieces on top of the columns show surpluses and shortages (i.e., how much the assigned number is above or below the authorized number). The figure shows a shortage of E7 soldiers coupled with a surplus of E6 soldiers. DL could help in this situation by training E6s sooner to increase the professional development level of the E7-eligible inventory.[7] Thus, as much as 30 percent of the E7 shortage could be filled by surpluses in the E6 grade of the same MOS.[8]

DL could also help with E7 shortages in 31P if there are instances of E6 fast-trackers who are already filling E7 positions. In that case, one might envision that soldiers would be allowed to continue to learn while on the job, but with the full support of schoolhouse instructors at a distance. Moreover, those soldiers, who will gain substantial experience while on the job, might also be allowed to test out of substantial portions of the course when they eventually take it.

POTENTIAL FORCEWIDE BENEFITS OF ACCELERATING TRAINING THROUGH DL

Accelerating BNCOC and ANCOC using DL can reduce shortages in trained personnel by increasing the professional development level of personnel in E6 and E7 positions. We estimate that 8,500 E6 and E7 soldiers could benefit from accelerated training through DL.

---

[7]Of course, actually increasing in the number of promotions to E7 is a decision that is independent of DL.

[8]The potential effect of accelerating ANCOC in the 31P case is limited to the surplus in the E6 grade. The fill rate for E7s was only 76 percent in June 1999, 44 NCOs short. If the training for fast-trackers in E6 could be accelerated, the overall E7 fill rate would improve, but only to the extent that surpluses existed at the E6 level. In the case of 31P, a surplus of 13 exists, 30 percent of the total E7 shortage.

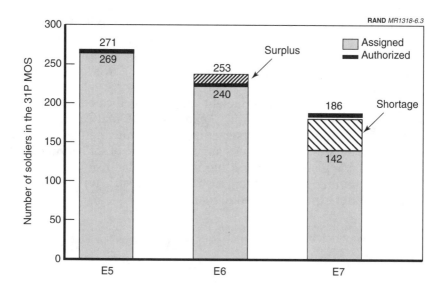

Figure 6.3—Using DL to Accelerate NCO Education Can Help Enable the Reduction of Grade-Level Shortages

Those soldiers fill positions representing 2.4 percent of all authorizations, and 8.9 percent of E6 and E7 authorizations. The majority were "fast-trackers" serving above their grade level without formal training. The remainder were soldiers who, although serving at grade level, had not yet received the training required for that grade. Although we did not examine BNCOC and ANCOC courses in this report, there may also be, as in the case of reclassification courses, opportunities to increase the cost-effectiveness of that training using DL. Finally, with the new NCO Educational System (NCOES) model projecting more individual training for NCOs, we think the level of potential application of DL for BNCOC and ANCOC will increase in the future.

However, the idea of using DL to accelerate training must be approached cautiously. Although a DL student need not wait for an opening in a residential training program, a poorly implemented DL program could easily not only fail to decrease the time to training completion, but actually increase it. This can occur because DL places more responsibility on the student and the chain of command in an environment with many competing demands. Moreover, because asynchronous portions of DL training will not have to be completed on a continuous basis, the time it takes a soldier to finish training can be longer than the course length. Lower completion rates and longer elapsed times to course completion can easily result in this environment, unless sufficient support is dedicated to achieve the desired training acceleration. Features of DL that can increase the speed of training include selecting highly motivated personnel for that type of training, fencing a large percentage of soldier time for training, using goal-setting and monitoring tools to support asynchronous training, and providing easy access to instructors and other subject-matter experts to expedite training.

Chapter Seven

CONCLUSIONS

We conclude that DL can contribute to strategies to reduce shortages and improve fill rates and can facilitate the Army's efforts to expand on those strategies (e.g., widening the window for reclassification). DL is particularly suited to making the reclassification, cross-training/MOS consolidation, and acceleration of training options more attractive to soldiers and commanders and more cost-effective for the Army. In addition, the three DL-based strategies will be useful in filling personnel gaps at both SL1 and NCO levels, and they will reduce the associated per-soldier cost of reducing shortages. Finally, the strategies can reduce the inevitable cost of force structure imbalances in a dynamic system, and they can indirectly improve the effectiveness of SRBs in reducing shortages. While the benefits identified in this analysis do not generally translate into current budget savings, we conclude that DL will increase the effectiveness of the overall process of reducing MOS shortages and will allow significant cuts in the Army's future cost of reducing personnel shortages.

However, realizing all these potential benefits requires careful implementing of the DL program. This means making earlier choices of courses for conversion during DL's long implementation period and concentrating on those courses that are most amenable to DL and that will help most to reduce the shortage problem (i.e., those focused on shortage MOSs, consolidating MOSs, and training problem MOSs). Most important, it also means creating DL courses that are attractive to students, commanders, and the Army, with sufficient flexibility to easily integrate into varying soldier career paths. In this regard, the DL program should emphasize the maximum use of emerging learning technologies to help reduce learning time (and,

thus, shorten courses) and to allow significant portions of the training to be done at home station. In addition, the DL program should
strive to avoid the past pitfalls in industry and academia by providing
sufficient student support to ensure speedy completion without
increased PERSTEMPO or course attrition, and to provide sufficient
administrative support for scheduling, monitoring, and recording
training results. Finally, DL needs to provide courses as modularized, "just-in-time" training to take full advantage of opportunities to
reduce unnecessary training and to allow refresher training on
demand.

The above list of specific DL characteristics for achieving the Army's
personnel readiness goals underscores the need for DCSPER, as well
as the Army as a whole, to work closely with the training community
to develop the kind of DL program that can maximize benefits in all
parts of the Army. In particular, we recommend that the Army
actively pursue the conversion of significant portions of some of its
longest reclassification courses to DL. Since the longer courses tend
to be the more technical ones, and since many technical skills are
amenable to DL, there is considerable potential for DL to help make
training in these skills easier to schedule, easier to complete, and
more efficient. Through a carefully designed monitoring program,
the Army's personnel and training communities would be able to
shed some additional light on the best ways to select material to be
presented using DL, the best use of the various DL technologies to
impart the required skills, and the best set of training support systems for ensuring that DL succeeds in the AC environment. Since
many of these skills and skill groups are found in other services and
in the private sector, continuing and sharing this work can provide
benefits that go well beyond the Army.

MEASURING THE EFFECTS OF DL-BASED AND NON-DL-BASED STRATEGIES ON SHORTAGES

The purpose of this appendix is to conceptually describe the methodology used to answer several key questions in the analysis. The questions our analysis needed to answer were the following:

1. How much will Army personnel inventory increase (at a grade level of detail) from adding 100 new soldiers per year in any of the following ways:

    a. as E3–4 reclassifications?

    b. as E5 reclassifications?

    c. as accessions?

2. How much will Army personnel inventory decrease (at a grade level of detail) from the removal of a standard reenlistment bonus?

Inventory Projection Models

To answer these questions, we used two steady-state Inventory Projection Models (IPMs). The reason for developing two models was the following: One IPM was developed for CMF 67 (aircraft maintenance), one of the occupational groups chosen for further study in the analysis. The other IPM used data for the force as a whole, and it was used to verify that the conclusions we drew about CMF 67 in this report would also apply to the force as a whole. We found that the conclusions reached in this report using IPMs did apply to the force as a whole, as well as to CMF 67.

As suggested by the questions posed above, the output of primary interest from the IPMs was "inventory by grade." The inputs, on the other hand, were more complex and numerous, and they are listed in Table A.1, along with the ultimate source of the baseline data used. A number of the inputs (on loss rates and promotion distributions, numbers 2, 3, and 4 in the table) came from out-year projections of the MOS Level System (MOSLS), a dynamic inventory projection model of the enlisted force at the MOS level of detail.[1] Other inputs (on accessions and promotion probabilities, numbers 1, 5, and 6 in the table) were outputs of a different kind of IPM used in another

Table A.1

IPM Input Data and Sources

Input data	Source
1. The number of accessions	Output of demand-pull IPM model*
2. Loss rates (i.e., percent of inventory that separates) by grade and year of service	Average of MOSLS projections, FY02–04, as of December 1997
3. End-of-term-of-service (ETS) loss rates by grade and year of service	MOSLS projection FY01, as of December 1998
4. Year-of-service distribution of pro-motions into each grade (percents summing to 100%, each grade)	Average of MOSLS projections, FY02–04, as of December 1997[†]
5. Promotion probabilities by grade (e.g., given promoted to E5, probability promoted to E6).	Output of demand-pull IPM model*
6. The total number of promotions by grade	Output of demand-pull IPM model*
7. The number of yearly reclassifications/ prior-service accessions by grade and year of service	Enlisted Master File (EMF) in FY99

*Described in Shukiar, Winkler, and Peters (2000).

[†]Adjusted in Shukiar, Winkler, and Peters (2000).

---

[1]MOSLS has been developed and maintained by GRC International, Inc., for the U.S. Army DCSPER's Military Strength Analysis and Forecasting Directorate. The model's mathematical logic is documented in *Documentation Updates for Mathematically Complex Programs in ELIM, MOSLS and OPALS,* September 1996, written by GRCI as part of the Strength Management Systems Redesign.

RAND analysis that dealt with 67 CMF. All these input data were made available to us in another RAND report (Shukiar, Winkler, and Peters, 2000), which displays baseline values for all the variables involved.

We could not directly use the models developed in the other RAND project because they were of a different type. Specifically, the models for the other RAND report used a so-called "demand-pull" methodology, which imposes inventory (i.e., the "demand") in each grade as an input.[2] The purpose of a demand-pull model is to explore what combination of promotion, accession, and other personnel policies would be required to support a given endstrength.

In this project, we required the development of a "supply-push" IPM, which has inventories as an output. The purpose of a supply-push IPM is to determine how inventory will change with changes in accessions and personnel policies. The concept here is that annual accessions (the "supply") are "pushed" through succeeding years of service by the outcomes of personnel policy to form a complete force structure. The personnel policies of interest in this project dealt with reclassification, prior-service accession, and standard reenlistment bonuses.

To understand the flows in an IPM model, consider a random cell in an inventory matrix by grade and year of service (Figure A.1). The shaded cell represents inventory at grade = g and year of service = Y (beginning of year), which we will represent as (g, Y). There are three flows into (g, Y) that determine its value: (1) inventory at (g, Y – 1) that continues in the same grade; plus (2) the number of soldiers entering from outside the occupation in that grade and year of service—either reclassifications from other MOSs or prior-service accessions; plus (3) the number of promotions in from cell (g – 1, Y – 1). Similarly, there are three flows out of (g, Y) that occur one year later: soldiers are either promoted to the next grade, continued in the grade without promotion, or separated from the Army.

---

[2]The basic mathematical formulation of demand-pull IPMs is fully described in Shukiar, Winkler, and Peters (2000).

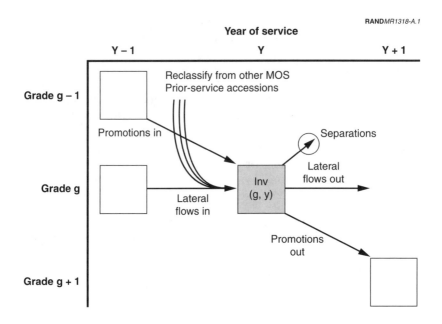

Figure A.1—IPMs from the Cell Level

Figure A.2 shows how the random cell in Figure A.1 fits into a total model structure. That figure shows all the possible inventory cells of the 67 CMF model using year of service 1–30 and grades E1–3 through E9. We have placed the random cell we examined earlier in position (E4, 3), and indicated its three inputs and outputs. Note that only some of the other cells have promotion inputs and outputs and lateral moves within the same grade. This is due to Army personnel policies that specify in what year of service promotion can occur and when soldiers are forced out of the Army if they have not reached a certain grade level.

How the IPMs Work: A Numerical Example

To understand how the supply-push IPMs worked in our analysis, we have constructed a simplified numerical analysis. Specifically, we show how 1,000 accessions would be "pushed" through the grade and year-of-service structure to determine inventory at cell (E4, 3).

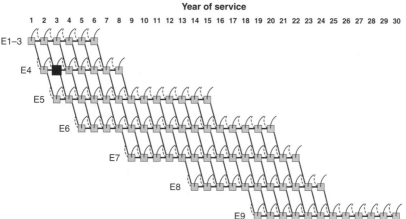

Figure A.2—Global IPM Structure and Flow

Then, to show the estimated effects of personnel policy, we show how that "baseline" inventory would change if 100 reclassifications were added to the accessions.

Table A.2 shows selected "baseline" inputs for the 67 CMF IPM, enough to calculate the inventory of interest at (E4, 3). The inputs include separation or loss rates, the promotion distribution, and promotion probability. We briefly explain each. First, loss rates are given for three years of service (YOS) and two grades. The loss rate of 17.3 percent at (E1–3, 1) means that, of all E1–3s in YOS 1, 17.3 percent will separate by the end of the year. Basically, the 17.3 percent rate at (E1–3, 1) represents "training" losses plus attrition during the remainder of the first year of service. The high loss rate of 34.8 percent at (E1–3, 3) indicates that more than a third of soldiers not promoted to E4 during the first two years of service will leave the Army after their third year.

Second, the promotion distribution is shown over three grades and four years of service. In the E1–3 column, all "promotions" enter E1–3 in YOS 1; that is, they are accessions. In the E4 column, 67 percent enter the grade at the beginning of YOS 2, and 33 percent enter the

Table A.2

Selected Baseline Inputs of CMF 67

Characteristic	E1–3	E4	E5
Loss rate			
YOS=1	17.3%		
YOS=2	10.2%	5.4%	
YOS=3	34.8%	7.2%	
Promotion distribution			
YOS=1	100.0%		
YOS=2		67.0%	
YOS=3		33.0%	18.5%
YOS=4			39.2%
Promotion probability	.768	.665	.415

grade at the beginning of YOS 3. Note that the percentages add to 100, indicating that all promotions to E4 in the model occur in the beginning of the second or third years of service. Under E5, the percentages shown do not add to 100, as some of the promotions occur after the fourth year of service and are not shown in the table.

Third, the promotion probability for each grade is given at the bottom of Table A.2. The number in the E4 column, for example, indicates that the probability that those reaching E4 grade will eventually be promoted to E5 is .665. In other words, given that a soldier was just promoted to E4, he or she has a .665 probability of being promoted to E5 sometime during his or her career. This also implies that he or she has a .335 probability of separating from the Army as an E4.

To see how the inputs of Table A.2 can determine inventories, we follow model computations one year of service at a time. Figure A.3 begins with the 1,000 accessions, which is both the input to and beginning inventory of cell (E1–3, 1). The right-hand side of the figure shows the flows out, that is, what happened to the 1,000 accessions at the end of the first year. Losses of 173 are determined by applying the appropriate loss rate (17.3 percent) to the cell inventory of 1,000. Promotions of 515 are determined by first determining total promotions in the grade (the .768 probability of promotion times the grade inventory of 1,000), then calculating promotions in that YOS (total

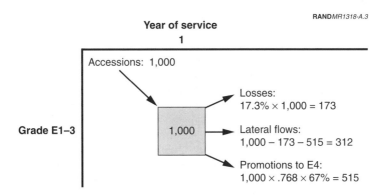

RAND*MR1318-A.3*

Figure A.3—Baseline Inventory and Flows of First Year of Service

promotions times the 67 percent of those promotions that occur at the end of the first—or beginning of the second—year of service). Finally, the lateral flows of 312 are calculated as a residual, that is, the original inventory of 1,000 less the 173 losses and the 515 promotions.

Figure A.4 shows inventories, flows in, and flows out for the second year of service. Note that the two inventory amounts in the second year of service are determined by flows out of the first year of service: namely, the 312 lateral flows in the first year of service become the E1–3 inventory in the second year, and the 515 promotions from the first year of service become the E4 inventory in the second year. At the end of the second year, inventories from each of the two second-year cells flow out as losses, promotions, and lateral flows in the same way as before. The right-hand side of Figure A.4 shows the calculations. Refer to Table A.2 to see the source of the data in the equations.

Figure A.5 shows inventories and input flows for the third year of service. It also shows the output flows for cell (E4, 3), our target inventory in this example. As before, the inventory amounts in the third year of service are determined by flows out of the second year of service. For example, in (E4, 3) the 646 inventory is determined by adding the promotions from (E1–3, 2) and the lateral flows from (E4, 2). To complete our baseline picture, Figure A.5 shows the out-

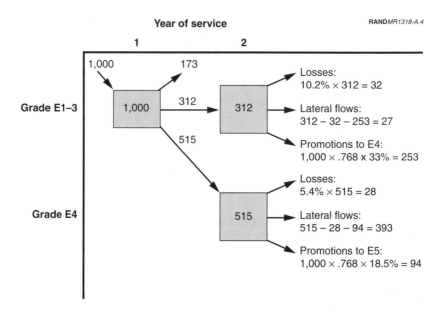

Figure A.4—Baseline Inventory and Flows for Second Year of Service

put flows and calculations for (E4, 3). While our example stops here, the model continues "pushing" the inventory from one year of service until the next, up to 30 years.

Once the baseline values are in place, the IPM can be used to estimate the effect of a policy change. For example, consider a decision to add to baseline inventory 100 yearly reclassifications from other MOSs at the E4 grade level. We suppose that 67 of the 100 reclassifications would flow into the occupation at the beginning of YOS 2, and the remaining 33 would come in at the beginning of YOS 3.

Figure A.6 shows the marginal effect on (E4, 2) inventory of the 67 reclassifications in the second year. First, the inventory at (E4, 2) would increase by 67. Second, at the end of the year, these 67 would either separate from the Army, be promoted to E5, or continue in E4. Figure A.6 shows the now-familiar calculations. Note that because we assume the reclassifications have the same characteristics as organic personnel in that grade and YOS, the inputs regarding loss rates, promotion probabilities, and promotion timing for the calcu-

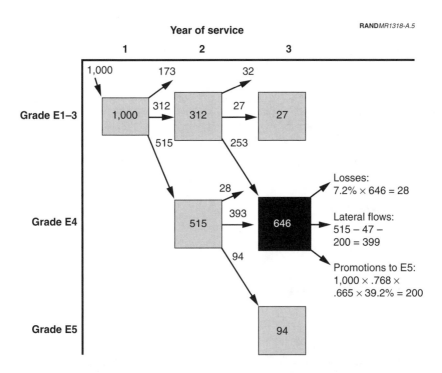

Figure A.5—Baseline Inventory and Flows for Third Year of Service

lations can be drawn directly from Table A.2; that is, once the 100 soldiers are reclassified into this CMF, they behave like soldiers already in the CMF for separation purposes, and they are subject to the same promotion probabilities as those already in the CMF.

Figure A.7 shows the marginal effect on inventories during the third year of service. Here, we see that cell (E4, 3) increases by 84 (due to the 51 lateral flows from the previous year, and the 33 additional re-classifications that come in for that year) and pushes 78 of those into YOS 4 (52 lateral flows and 26 promotions to E5). In subsequent grades and years of service, the marginal effect of the reclassifications is further increased. Note that considering only three years of service and two grades, the 100 reclassifications have already increased force size by nearly 250 personnel (67 + 84 + 12 + 52 + 26 = 241).

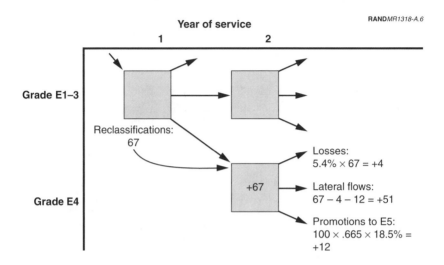

Figure A.6—Marginal Effect of Reclassifications in Second Year of Service

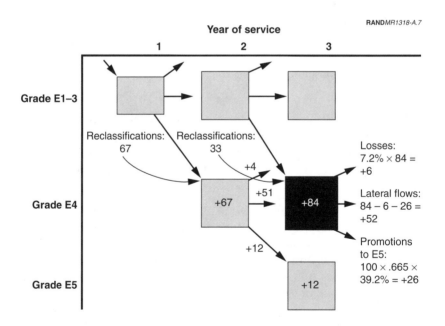

Figure A.7—Marginal Effect of Reclassifications in Third Year of Service

LENGTH OF RECLASSIFICATION COURSES UNDER DISTANCE LEARNING

This appendix provides further discussion of our assumption that DL reclassification courses in the Army can be 30 percent shorter than their corresponding AIT counterparts.

The 30 percent number originated from research conducted by Orlansky and String (1979), who examined the results of some 30 studies of the effects of DL on military training. The 30 percent figure was the median effect on course length for the courses they examined. In addition, it should be noted that the variance of the effect of DL on course length was large; in fact, three of the courses studied by Orlansky and String actually required more time after conversion. In explaining the overall reduction of course length across studies, Orlansky and String point out that one likely reason for this effect is that self-paced DL instruction allows students to spend only as much time as needed to achieve a given performance standard.

The Army Science Board, in its 1997 study of Army DL, also concluded that course length could be reduced 30 percent. In making its determination, the board cited research demonstrating that the application of modern learning technology can lead to large reductions in learning time. Perhaps more important, the board cited reasons for course reduction having to do with the conversion process. Independent of learning technology and media, course developers in academic settings have found that when an existing RL class has been examined for critical tasks, and then modified to meet these objectives, significant reductions in course length have been achieved (Army Science Board, 1997, p. 15).

While experience in academia does not necessarily translate to the Army, reclassification courses are more likely to be successfully shortened than other types of Army courses. The reason has to do with differences in student populations. Students in reclassification courses already have, by definition, considerable Army-specific experience and success. As a result, reclassification students are more likely to learn faster than the new recruits in AIT courses, who are trying to learn not only a skill, but also how to succeed in the Army.

To test the validity of the 30 percent assumption in the context of current Army training, we examined the eight reclassification courses scheduled to begin conversion in FY98 (see Table B.1). At the beginning of FY00, the designers of these courses were forecasting a 27 percent reduction in course length, a number nearly as large as the 30 percent assumption. It should be noted that a large variance existed around the average; some courses did not reduce in length at all, while others reduced by as much as 41 percent.

Table B.1

DL Redesign Effect on Length of Reclassification Courses: Early Results

CMF	Course	Course Name	Length in Weeks		Percent Reduction
			FY98 AIT Resident*	DL Redesigned**	
Ordnance	63B10	Light Wheeled Vehicle Mechanic	10.0	6.0	40%
Quartermaster	57E10	Laundry & Show Specialist	4.0	3.5	13%
Aviation Logistics	657T10	UH-60 Helicopter Repairer	14.6	8.6	41%
Engineer	51B10	Carpentry/Masonry Specialist	7.4	7.4	0%
Field Artillery	13P10	MLRS Fire Direction Specialist	14.6	8.6	41%
Military Intelligence	96R10	Ground Surveillance Systems Operator	5.4	4.8	11%
Finance	73C10	Finance Specialist	7.0	5.8	17%
Adjustment General	71L10	Administrative Specialist	5.0	5.0	0%
Total			68.0	49.7	27%

*From FY98 ATRRS listing and data presented at TADLP Program Manager's Annual IPR, September 1999.

**Data presented at TADLP Program Manager's Annual IPR, September 1999.

REFERENCES

Army Science Board, Office of the Assistant Secretary of the Army (Research, Development, and Acquisition), *Distance Learning*, 1997 Summer Study Final Report, December 1997.

Barry, M., and Runyan, G.B., "A Review of Distance Learning Studies in the U.S. Military," *The American Journal of Distance Education*, Vol. 9, No. 3 (1995), pp. 37–47.

Farris, Hilary, William L. Spencer, John D. Winkler, and James P. Kahan, *Computer-Based Training of Cannon Fire Direction Specialists*, Santa Monica, CA: RAND, MR-120-A, 1993.

Jones, Maj. Gen. Anthony, "Aviation Modernization Strategy: 2000 and Beyond," *Army Aviation*, May 31, 2000.

Keene, S. Delane, and James S. Cary, "Effectiveness of Distance Education Approach to U.S. Army Reserve Component Training," in Michael G. Moore (ed.), *Distance Education for Corporate and Military Training*, University Park, PA: Pennsylvania State University, 1992.

Leonard, Henry A., John D. Winkler, Anders Hove, Emile Ettedgui, Michael Shanley, and Jerry Sollinger, *Enhancing Stability and Professional Development Using Distance Learning*, Santa Monica, CA: RAND, MR-1317-A, 2001.

Hix, W. Michael, Herbert J. Shukiar, Janet M. Hanley, Richard J. Kaplan, Jennifer H. Kawata, Grant N. Marshall, and Peter J.E. Stan, *Personnel Turbulence: The Policy Determinants of Permanent*

Change of Station Moves, Santa Monica, CA: RAND, MR-938-A, 1998.

Orlansky, Jesse, "Effectiveness of CAI: A Different Finding," *Electronic Learning,* Vol. 3, No. 1, September 1983.

Orlansky, Jesse, and Joseph String, *Cost-Effectiveness of Computer-Based Instruction in Military Training,* Institute for Defense Analyses, IDA Paper P-1375, April 1979.

Orlansky, Jesse, and Joseph String, "Computer-Based Instruction for Military Training," *Defense Management Journal,* Second Quarter 1981, pp. 46–54.

Phelps, Ruth H., Rosalie A. Wells, Robert L. Ashworth, Jr., and Heidi A. Hahn, "Effectiveness and Costs of Distance Education Using Computer-Mediated Communication," in Michael G. Moore (ed.), *Distance Education for Corporate and Military Training,* University Park, PA: Pennsylvania State University, 1992.

Polich, J. Michael, Bruce R. Orvis, and W. Michael Hix, *Small Deployments, Big Problems,* Santa Monica, CA: RAND, IP-197, 2000.

Program Management Office, TADLP, "The Army Distance Learning Program Economic Analysis," September 13, 2000.

Rumble, Greville, *The Costs and Economics of Open and Distance Learning,* London: Kogan Page Ltd., 1997.

Shukiar, Herbert, John D. Winkler, and John E. Peters, *Enhancing the Retention of Army Noncommissioned Officers,* Santa Monica, CA: RAND, MR-1186-A, 2000.

"Soldier by Day, Online Student by Night," *Los Angeles Times,* July 10, 2000, p. A1.

Wild, William G., Jr., and Bruce R. Orvis, *Design of Field-Based Crosstraining Programs and Implications for Readiness,* Santa Monica, CA: RAND, R-4242-A, 1993.

Winkler, John D., and J. Michael Polich, *Effectiveness of Interactive Videodisc in Army Communications Training,* Santa Monica, CA: RAND, R-3848-FMP, 1990.

Winkler, John D., et al., *Future Leader Development of Army Non-commissioned Officers: Workshop Results,* Santa Monica, CA: RAND, CF-138-A, 1998.